INTERCONNECTIONS FOR COMPUTER COMMUNICATIONS AND PACKET NETWORKS

INTERCONNECTIONS FOR
COMPUTER COMMUNICATIONS AND PACKET NETWORKS

ROBERTO ROJAS-CESSA
NJIT, Newark, New Jersey

CRC Press
Taylor & Francis Group
Boca Raton   London   New York

CRC Press is an imprint of the
Taylor & Francis Group, an **informa** business

CRC Press
Taylor & Francis Group
6000 Broken Sound Parkway NW, Suite 300
Boca Raton, FL 33487-2742

© 2017 by Taylor & Francis Group, LLC
CRC Press is an imprint of Taylor & Francis Group, an Informa business

No claim to original U.S. Government works

Printed on acid-free paper
Version Date: 20160712

International Standard Book Number-13: 978-1-4822-2696-6 (Hardback)

Visit the Taylor & Francis Web site at
http://www.taylorandfrancis.com

and the CRC Press Web site at
http://www.crcpress.com

To Ad, Marco, and Nelli,
who make every day special.
To my parents, brothers, and sisters.

Contents

III   Data-Center Networks   239

Preface

It is easy to find that most of the information we consume and that systems and machines use for their proper operation is being digitized and packetized. These bits and packets travel through the webs that interconnect them. These webs are an important component of these communications systems and determine in great measure how fast and how much information travels. So, it is no surprise that interconnection networks play an important role in almost every communications system. As we move into a software-defined communications systems, interconnections are expected to play an even increasingly larger role in these systems.

The idea of writing this book comes from students requesting accessible materials that would introduce them to the different topologies and algorithms that are the basis for building, understanding, and designing interconnection networks, in a straightforward and approachable manner. Some of the existing books are comprehensive and present some research-oriented information. However, the amount of information and the speed in which the Internet helps to generate even more of it may confuse the readers, especially those who are about to be initiated in the field, those who may seek a selective collection of works to grab a general understanding of the field in an expedited manner, or those who may need to identify a specific research challenge. The material covered in this book and the discussion provided attempts to fulfill those objectives.

The intended audience for this book are those individuals with a good sense of curiosity. Although the writing of this book has been motivated by the needs of graduate students, the material may not require strict prerequisites. Some knowledge in computer architecture, computer networking, Internet protocols, and introductory probability may be helpful. The book can be used as a textbook for advanced undergraduate and graduate courses, or for independent and individual learning.

Each chapter also presents some exercises to help the reader to reinforce and clarify the discussions presented throughout the book. These exercises are of basic level but instructors can use them as a starting point and get into more complex exercises.

Organization of the Book

The book is organized into three parts: interconnection networks for multiprocessors, packet networks, and data centers. The first part is dedicated to the basics of communications for multiprocessor systems, and the different interconnection networks used in this field. The larger portion of the book discusses interconnection networks for packet switching, and the third part, a brief introduction to data center interconnection networks. Each of the three parts of the book can be read separately but consideration of the three parts may enhance the understanding of how interconnection networks work and what their design philosophies are, and it may also provide the reader with a wider scope of applicability of these networks.

These three parts have been put together to offer an overall view on how interconnection networks operate and their design key points. The book, however, discusses packet networks in much more depth than the two other topics. The first part of the book introduces the conventional and well-known interconnection networks for multiprocessor systems and some of the routing mechanisms used in them (Chapters 1 and 2). These routing mechanisms exploit the parallelism, symmetry, and modularity of these interconnections.

In the second part of the book, Chapter 3 presents IP lookup as an introduction to packet switching. This chapter discusses the role of memory in building forwarding tables and schemes to make IP lookup more efficient. Chapter 4 discusses packet classification, which is a topic very much used in filtering and identifying packets in almost every network in the Internet.

Chapter 5 covers the basics of packet switching, to introduce the reader with the terms and metrics used to analyze the performance of packet switches.

Chapter 6 presents input-queued packet switches, which are a very popular and practical architecture in network equipment manufacturers.

Chapter 7 discusses shared-memory switches, which uses memory as part of the switching system (or fabric). The chapter discusses several practical strategies to minimize the use of memory speedup and also some techniques used to avoid memory hogging by misbehaving users.

Chapter 8 discusses internally buffered packet switches. This switch architecture, as in shared-memory switches have memory in the switch fabric, show many of its advantages over other switch architectures and also show that it can be used to relax timing constraints for the transmission of high data rates.

Chapter 9 presents load-balanced packet switches. These packet switches are among the most recently developed. This chapter reviews these architectures and highlights the remaining challenges to make them a reality.

Chapter 10 presents Clos-network switches. Different from the other switch architectures, Clos networks are used for building very large switches as the other architectures cannot scale up.

In Part III of the book, Chapter 12 presents different interconnection networks for data centers. It also discusses the different performance metrics used in the application of interconnection networks on data centers.

Suggested Coverage

The complete material in this book may be used for a graduate course covering most of the three parts, or may be partitioned as desired. The chapters may be discussed separately except for the sequence of Chapters 2 and 3. Also, it is suggested to cover the basics of packet switching, Chapter 5, before covering any other packet-switching chapter for a more efficient understanding. Some of the chapters of this book can also be used for an undergraduate course. It is highly suggested that undergraduate students have some knowledge of computer architecture for covering Part I or computer networks for Part II of this book.

The chapters of IP lookup and packet classification have been placed before those about packet switching as these two chapters can be discussed as standing-alone chapters. Although they are part of packet switching, they include information from a higher layer of the Internet protocol stack, while the packet-switching chapters can be placed in the lower layers of the same protocol stack.

Comments or suggestions are welcome. Feel free to send them to rojas@njit.edu.

Acknowledgments

This book would not be possible without the direct or indirect support of many people. First, I would like to acknowledge the many students who have taken my class and have requested materials that directly introduce them to the topics so they can get ahead faster and effectively. I would also like to thank Nora Konopka (Taylor & Francis) for inviting me to jump into this project, assistance provided along the way, and helping me with keeping up with deadlines. Special thanks are given to Prof. Jonathan H. Chao (New York University) for introducing me to this field and providing support during all this time.

Some of the materials presented in this book have been part of research work with Prof. Jonathan H. Chao (New York Univerity), Dr. Zhen Guo, Prof. Eiji Oki (The University of Electro-Communications), Prof. Ziqian Dong (New York Institute of Technology), Prof. Chuan-bi Lin (Chaoyang University of Technology), Prof. Taweesak Kijkanjanarat (Thammasat University), and especially, with several of my students. I would like to recognize and thank Yagiz Kaymak, Vinitmadhukar Sahasrabudhe, and Prof. Ziqian Dong for assistance with some of the chapters of the book. I would like to thank Yao Wang for providing initial help with artwork. Special thanks go to Prof. Hyesook Lim (Ewha Womans University) for contributing with some of the material on IP lookup and being supportive in the writing of this book. I would like to thank Prof. Ziqian Dong, Yagiz Kaymak, Sina Fathi Kazerooni, and Jingya Liu for reading early drafts. Finally, I would like to thank my students from the ECE681 "High Performance Routers and Switches" class, in the spring 2016 semester, for providing invaluable feedback and testing some of the exercises and examples.

Part I

Processor Interconnections

1

Multiprocessor Interconnection Networks

CONTENTS

Processors communicate with other processors, memory, and other shared resources for fast computations. These communications take place over an interconnection network. The performance of the communications and, in turn, the computations made by the processors, as well as the scalability of these multiprocessor systems, depend on the characteristics of the interconnection network used. This chapter introduces some of the interconnection networks used by multiprocessor systems and their properties.

1.1   Introduction

Parallel computer systems, often referred to as multiprocessor systems, form a class of computing devices that may contain a minimum of two, but often more and distinct central processing units (CPUs) within the same logical entity. Historically, the motivation to build such systems was to achieve high computing performance to solve numerous industrial and scientific problems that demand extensive computations. Such applications can be quite varied and complex and could be an overwhelming load for stand-alone systems using a single processor. Examples of such applications may be found in aerodynamic simulations, air traffic control, air defense systems, biomedical image processing, chemical reaction simulations, weather forecasting, deoxyribonucleic acid (DNA) sequencing and others [14]. Parallel and distinct processing elements may be adopted to overcome the physical characteristics (e.g., processor clock speeds) of a specific computing system or processor design.

The applications (or programs) targeted to run on parallel systems are associated to algorithms with high computational complexity, large data sets, and stringent time constraints [156]. For example, the output of a 24-hour weather forecasting model is of no use if it takes longer than the stipulated time (24 hours) to process data and estimate a reasonable forecast [156].

Parallel multiprocessor systems initially found their application in super computers, mainframes, and super mini systems. The ability to complete floating point arithmetic operations, defined as floating point operations per second (FLOPS), is often used as a measure of the performance achieved by such systems. The delivered performance of such systems largely depends on the computing algorithm (and its complexity) being executed, the topology of the interconnect used to communicate the different processing elements (PEs), and other factors. For example, the IBM Blue Gene project, unveiled in 2007, was reported to deliver a performance of 13.6 Giga-FLOPS and it is built using four Power PC cores [6]. To achieve this level of performance, all PEs must work cooperatively to complete a single (or multiple) task(s). Parallel processing systems rely on simultaneous distributed operation of programs, jobs, and subroutines to complete an assigned task [178]. Each processing unit therefore must rely on a fast reliable transport mechanism to communicate across multiple processing elements. In the context of parallel multiprocessor systems, this mechanism is generally referred to as a multiprocessor interconnection network or simply the interconnection network (IN).

1.1.1 Multiprocessor Computing Architectures

Large-scale parallel systems allow a variety of ways to organize the processors, memories, and interconnection networks used to build such systems. The selection of those components define the characteristics of parallel multiprocessor systems.

We briefly introduce the working principles of single instruction multiple data (SIMD) and multiple instruction multiple data (MIMD) stream architectures, often used in multiprocessor systems.

1.1.1.1 SIMD Machines

A model of an SIMD machine consists of a control unit, N processors, N memory modules, and an interconnection network. The control unit broadcasts instructions to the processors and all active processors execute the same instruction at the same time, as a single instruction stream. Each active processor executes the instruction on data in its associated memory module. Thus there are multiple data streams. The interconnection network, sometimes referred to as alignment or permutation network, provides a means of communications between the processors and memory modules. An example of SIMD machines constructed based on the SIMD architecture is the STARAN computer [155].

1.1.1.2 Multiple SIMD Machines

A variation of the SIMD model that may allow a more efficient use of processors and memories is the multiple-SIMD machine. This machine is a parallel processing system that can be dynamically reconfigured to operate as one or more independent SIMD submachines of various sizes. A multiple-SIMD system consists of N processors, N memory modules, an interconnection network, and C control units, where $C < N$. Each of the multiple control units can be connected to a disjoint subset of processors, which communicate over an interconnection network. An example of such a multiple-SIMD is the Connection Machine CM 2 [155].

1.1.1.3 MIMD Machines

Differently from the SIMD machines, where all processors follow a single instruction stream, each processor in a parallel machine may follow a different instruction stream, forming an MIMD machine. The organization of an MIMD machine may consist of N processors, N memory modules, and an interconnection network. In such a machine, the IN provides communications among the processors and memory modules to run multiple instruction streams and multiple data streams. While in an SIMD system all active processors use the interconnection network at the same time (i.e., synchronously) in an MIMD system, because each processor executes its own program and inputs to the network arrive independently (i.e., asynchronously). Examples of large MIMD systems are the BBN Butterfly, the Intel iPSC cube, and the N Cube [155].

1.1.2   Multiprocessor vs. Multicomputer Systems

General-purpose large-scale parallel and distributed computer systems form two categories of systems with multiple processors: multiprocessors and multicomputers. The difference between these systems lies in the level at which interactions between the processors occur. A multiprocessor permits all processors to directly share the main memory and the processors address a common main memory space. In a multicomputer, however, each processor has its own memory space, and sharing of memory between processors occurs at a higher level, as with a complete file or data set. A processor cannot directly access another processor's local memory [14]. This difference defines the function of the underlying IN for each of the systems. Multiprocessor communication uses message passing and multicomputer communication exchanges large volumes of data in the form of files.

Multiprocessor systems may be classified into two major classes according to how memory is distributed. In a class of these multiprocessor systems, the main memory is situated at a central location so that the access time from any processor to the memory is similar. In addition to this central memory, which is also called main memory, shared memory, or global memory, each processor might host a local memory or cache. In the other class, the main memory is partitioned and attached to the processors, and the processors share the same memory address space. In such systems, a processor may directly address a remote memory, but the access time may be much larger than the access time to a local memory. As a result, partitioning and allocation of program segments and data play a crucial role in the overall performance of an application program [14]. In either case, an IN is needed to interconnect both systems but with different goals. It must be noted that the IN also interconnects other input/output resources across all nodes in the system. Herein, the term node generically refers to any logical endpoint connected on the interconnect network.

1.1.2.1   Need for an Interconnection Network

The use of a network for interconnecting multiple devices permits sharing resources through passing messages across multiple nodes. The performance demands on the interconnection networks are governed by the architecture of the system where processing units contend to access the shared resources and to pass messages in the case of tightly coupled systems. The topology of an interconnection network determines the characteristics and, possibly, the achievable performance. Topology is the pattern in which individual nodes of a network are interconnected to others. In this case, a node may be a switch (to interconnect one processor to any of multiple memories), processor, or memory.

Interconnection networks for parallel systems may have different design goals. A major difference among them is the distance between any two nodes in a parallel system. Conventionally all nodes in a parallel-system interconnection network would be treated with equivalent priority and permitted to exchange the equivalent amounts of data.

1.1.3 Topology of Interconnect Networks

The topology of interconnection networks for parallel systems may be coarsely divided into two classes: direct and indirect. In direct networks, each switch has a direct link to a processing node. In indirect networks, many switches in the network may be attached only to other switches [12]. Therefore messages being exchanged from one processing unit to another in indirect interconnect networks may have to pass through several switches before they arrive at the destinations.

1.2 Direct Networks

Direct networks consist of links that connect the nodes in a parallel computer [178]. Each node is connected to one or more of those interconnection links. Because the network consists of only links, nodes make the routing decisions. Dedicated hardware may be used in each node to select one of the interconnection links for forwarding a message toward its destination. Because a node is normally not directly connected to all other nodes in a parallel computer, a message transfer from a source to a destination node may be required to pass intermediate nodes to reach the destination node. Every node in the system typically consists of one or more processors (some of them may be associated local memory) and a hardware switch that controls the routing of messages through the node. Figure 1.1 shows an example of a direct network, a bi-dimensional (2D) mesh interconnection network. The nodes in this network have a direct and dedicated link to other nodes. Note that a node is not directly connected to all other nodes in the network.

The topology of a network determines some properties of direct interconnection networks. These are summarized below.

- Switch degree: number of outgoing links from a switch in a network.

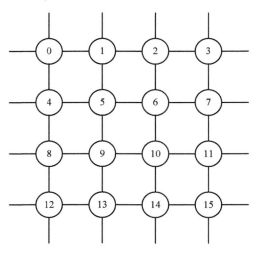

FIGURE 1.1
Example of a direct network: a 2D mesh with 16 nodes.

- Diameter: number of links crossed between nodes on maximum shortest path.

- Average distance: the number of hops to a random destination.

- Bisection: minimum number of links that, if removed, would separate the network into two parts.

1.2.1 Mesh Interconnect Networks

A basic topology of an interconnection network is the mesh network, as shown in Figure 1.1. This 2D mesh network has a node degree of four (i.e., four nodes per dimension, e.g., x and y dimensions, and a total of 16 nodes). Each node in the network is both a destination and switching node. As the figure shows, each of the four links of a node is interconnected to a neighbor, except for the edge nodes, which have a degree of three. In a mesh network, the connection between each pair of adjacent nodes may be unidirectional or bidirectional, with the bidirectional connections being considered as two unidirectional links. The 2D mesh network has a diameter of $2\sqrt{n}$ and a bisection bandwidth of $\sqrt{n}$, where n is the number of nodes in the interconnection network.

A mesh network is well suited for SIMD architectures because they match some data parallel applications (e.g., image processing, weather forecasting). Some advantages of a mesh network are short links that lead to use of a small-area footprint, having a regular layout that helps to simplify the design. A major disadvantage is that messages have to travel through a large number

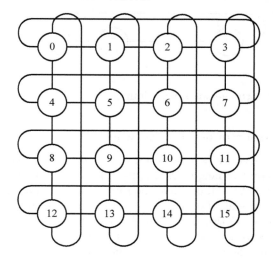

FIGURE 1.2

Example of a 2D torus network with 16 nodes.

of hops to reach a destination node (proportional to N for a network with N nodes). Each node contributes to latency and is a potential point of contention. The network size (in number of nodes) directly impacts some of the functions in the IN, such as buffering, transmission, and energy consumption.

1.2.1.1 Torus Network

A torus network is an extension of a 2D mesh network. Figure 1.2 shows an example of a torus network, where the network is a 2D mesh with all nodes with a degree of four. The nodes at the edge are then interconnected. This interconnect is called a torus because it looks as if a tin were wrapped around the nodes and edges such that it would form the shape of a doughnut. The diameter of a torus network is half that of a 2D mesh with the same number of nodes [183]. The connections at the boundaries not only reduce the diameter but also provide symmetry to the network. A torus enables full connectivity and increases path availability through redundancy. The added redundancy helps avoid contention for links and improves system performance.

The Blue Gene/L (BG/L) is a 64k (65,536) node supercomputer built by IBM. It uses a three-dimensional (3D) torus as the primary network for point-to-point messaging. In this machine, the nodes are arranged in a 3D cubic grid in which each node is connected to its six nearest neighbors with high-speed dedicated links. A torus was chosen because it provides high-bandwidth nearest-neighbor connectivity while eliminating edges. This topology yields a cost-effective interconnect that is also scalable [6].

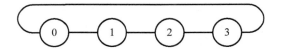

FIGURE 1.3
Example of a 1D-torus network with four nodes.

1.2.1.2   Ring Interconnection Network (1D Torus)

The N-node ring network is a variation of the torus interconnect. Each node is connected to only two neighbors. Figure 1.3 shows a ring network with four nodes, where three of the links have the same length and all nodes are placed in a linear arrangement. The last link joining node 0 and 3 is the longest as it wraps around the two nodes. In this IN, the data may be moved from one processor to another by a sequence of cyclic shifts. This IN has a large diameter (assuming bidirectional links) and its communication performance decreases as the number of nodes increases.

1.2.2   Honeycomb Networks

1.2.2.1   Honeycomb Mesh Network

The honeycomb mesh (HM) network uses a hexagonal interlocking pattern much like a bee's honeycomb and from which it derives its name. The honeycomb mesh has a tiled symmetry as shown in Figure 1.4 [165]. This HM is built as follows: one hexagon is a honeycomb mesh of size 1 with six PEs as nodes, and it is denoted as HM1. The honeycomb mesh HM2, or of size 2, is built by adding six hexagons to the boundary edges of HM1. Inductively, honeycomb mesh HMt of size t is obtained by adding a layer of hexagons around the boundary of HM$(t-1)$. The size t of HMt is determined as the number of hexagons between the center and boundary (inclusive). The number of nodes and links for an HMt network is $6t^2$ and $9t^2 - 3t$, respectively.

The HM network with n nodes has degree three and diameter of $1.63 \times \sqrt{(n)} - 1$. The network cost, which is defined as the product of degree and diameter, is $4.9 \times \sqrt{n}$. Honeycomb meshes have a 25% smaller degree and 18.5% smaller diameter than a mesh network with the same number of nodes. This is an important advantage of honeycomb meshes over square meshes. The advantage can be explained as follows. It is clear from Figure 1.4, which also shows that honeycomb mesh can be embedded onto the square grid, that the distance between two nodes remains virtually the same if a quarter of the edges is eliminated from the grid. Furthermore, by making a hexagonal rather than square boundary, the boundary is closer to a circular shape, thus reducing its diameter. The cost of a honeycomb network is about 39% less than the cost of a mesh-connected computer [165].

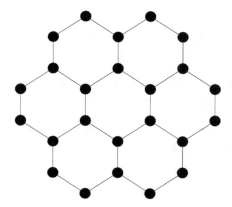

FIGURE 1.4
Example of a honeycomb mesh interconnect.

1.2.2.2 Honeycomb Tori Network

A variation of the honeycomb mesh is the honeycomb tori (HT). This IN is built by joining node pairs of degree two (i.e., their unused ports) of the honeycomb mesh. In order to achieve edge and vertex symmetry, the best choice for wrapping around seems to be the pairs of nodes that are mirrored symmetrical with respect to lines passing through the center of hexagonal mesh and normal to each of the edge orientations. Figure 1.5 shows a layout of an HT network.

The number of links in HTt is $9t^2$. The diameter of honeycomb torus HTt is $2t$, which is approximately twice as small as the diameter of a corresponding honeycomb mesh. This is the same ratio as in the case of a tori that expands mesh-connected nodes and hexagonal meshes [165]. Therefore, the cost of the tori is two times less than the cost of corresponding meshes.

In summary, the degree of three of honeycomb networks limits the construction of edge-disjoint paths from any two nodes to three. Maximizing the number of such paths is important for building fault-tolerant applications. It has been observed that this network has three disjoint paths between any two nodes (or degree three). Also, the bisection width of honeycomb torus is larger than that of a square torus with the same number of nodes. Figure 1.6 shows a hexagonal mesh of size 3 for reference.

1.2.3 Hypercube (Binary n-Cube)

A binary n-cube or hypercube network is a network with $2n$ nodes arranged with the vertices as an n-dimensional cube. A hypercube is considered as a generalization of a cube. To build a hypercube network, start with considering a single node, or node 0. After that, add another node placed a unit distance

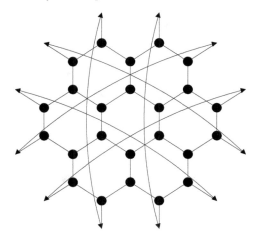

FIGURE 1.5
Layout of a honeycomb tori network.

away. This is called 1-cube. Figure 1.7 shows a 1-cube, 2-cube, and 3-cube to provide an intuitive idea of the relationship of a dimension and the construction of an n-dimensional cube. As the replication process continues further into an orthogonal direction, connecting all adjacent nodes, and updating the labels, the dimension of the network size increases and the number of bits used to identify a node also increases. Furthermore, the most significant bit (MSB) changes from 0 to 1 as we move across the plane. For an n-dimensional hypercube, each node label is represented by n bits, and each node has exactly n links [183]. Note that the labels of two nodes differ by exactly one bit value if these nodes are interconnected by a direct link. After selecting a bit position out of those in the label, the nodes of the n-cube can be divided into the 0-plane (i.e., all the nodes whose selected bit is 0) and the 1-plane. Every node in the 0-plane is attached to exactly one node in the 1-plane by one edge. There are 2^{n-1} such pairs of nodes, and hence 2^{n-1} edges. Figure 1.8 shows a 4D hypercube.

In hypercube network with 2^n nodes, each node has a degree of n, which also is the dimension of the hypercube [183]. In this network, two nodes of are directly connected if their Hamming distance is one (the node labels differ exactly by one bit). The Hamming distance of two binary numbers is defined as the number of bits the two numbers differ. The number of hops that a message traverses is equal to the Hamming distance from the source node to the destination node. The diameter of an n-dimensional hypercube is n. The bisection width is 2^{n-1}.

Hypercube networks are complex to scale; the degree of each node is incremented in each increased step and that also increases the dimension of the network. Therefore, the next larger hypercube has twice the number of nodes

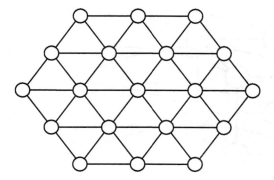

FIGURE 1.6
Hexagonal mesh IN of size 3.

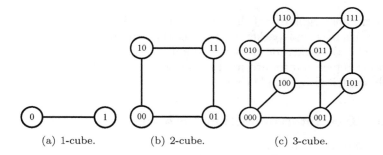

(a) 1-cube.   (b) 2-cube.   (c) 3-cube.

FIGURE 1.7
Hypercubes or binary (a) 1-cube, (b) 2-cube, and (c) 3-cube.

of the previous hypercube. The maximum edge length increases as network size increases. However, the small diameter and large bisection width make it an attractive network.

1.2.4 k-Ary n-Cube

All the above topologies discussed so far are characterized by having their network links arranged in several orthogonal dimensions, in a regular fashion. In fact, the k-ary n-cube class of topologies includes the ring, mesh, tori, and cube among others. The k-ary n-cube network is used to interconnect $N = k^n$ nodes, where n is equal to the number of dimension of the network, and k is the network radix, which is equal to the number of nodes in each dimension. For example, a k-ary 1-cube is equivalent to a k-node ring network (i.e., a network with k nodes which would scale in only one direction to form a ring), similarly, a k-ary 2-cube is equivalent to a k-node 2D mesh network or a torus network if the edge nodes circle back and connect together forming a loop. The

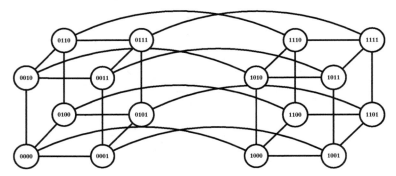

FIGURE 1.8
4D hypercube.

symmetry and regularity of these topologies simplify network implementation (i.e., packaging) and packet routing as the movement of a packet along a given network dimension does not modify the number of remaining hops in any other dimension toward its destination. This topological property can be readily exploited by simple routing algorithms.

1.2.5   Tree Interconnection Networks

At its simplest, a tree consists of a parent node and children or descendant nodes of that parent. The tree starts with a root node connected to its descendants and thus propagates downwards. The set of descendants could be possibly connected to a disjoint set of nodes. The descendant node that has no children is called a leaf node. Figure 1.9 shows the basic structure of a tree as a graph. This graph is known as a binary tree. As the figure shows, every node has a single parent and no loops. A tree with all leaf nodes having the same distance to the root node is classified as a balanced tree. Otherwise, the tree is classified as unbalanced. A tree in which every node but the leaves has a fixed number of descendants k is a k-ary tree.

Typically, trees are employed as indirect networks with processing nodes, as the leaves and all intermediate nodes, including the root node, are implemented as switches. A binary tree with N leaves has a diameter of $2 \log_2 N$. In this case, the binary address of each node is a $d = \log N$ bit vector, which specifies a path from the root to the destination. Figure 1.10 shows a binary tree where all intermediate nodes are switches and the leaves are processors. In this tree, the switches only forward data from processors to other processors (or one port to another, according to the data destination). Switches use the bit vector to identify to which link they should forward the data. Other architectures may use processors for all nodes in the network.

For a balanced tree, each node with the exception of the root and the leaf nodes has a node degree of three. Leaf nodes are connected to only their

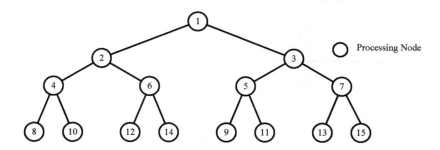

FIGURE 1.9
Generic tree interconnect.

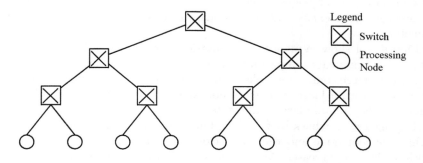

FIGURE 1.10
Tree with switches as intermediate nodes and processing nodes as leaves.

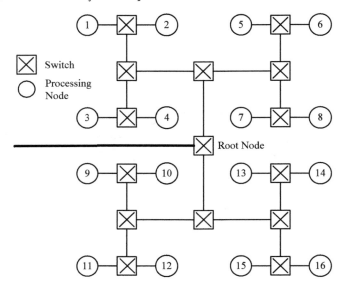

FIGURE 1.11
A 2D planar layout of a tree interconnect.

parents and the root node in a balanced network is independent and not part of a larger interconnect. The bisection width is one as it is possible to split the tree into two sets differing by at most one node in size by deleting either edge incident to the root. Therefore, trees are well suited for algorithms with localized communication needs. This interconnect provides a unique route between any pair of nodes. However, fault tolerance is compromised since no alternate path exists and the average distance is almost as large as the diameter.

Figure 1.11 shows a planar representation of the three in Figure 1.10. The flat tree is a practical example on the implementation of a binary tree, without recurring to large connection links between the root and its children. However, large trees (with a large number of nodes) may be complex to build in this fashion as they require large real estate.

The biggest drawback by far of the binary tree as an interconnect network is that the root node and the nodes close to it become a bottleneck as the depth of the tree increases. An obvious method to overcome this drawback is a provision of additional bandwidth as we get closer to the root node.

1.2.5.1   Hyper Tree

The hyper tree is a network topology that aims to combine the best features of expansible tree structures and the compact n-dimensional hypercube. The two underlying structures permit two distinct logical views of the system. Thus, issues that map onto a tree structure can take advantage of the binary

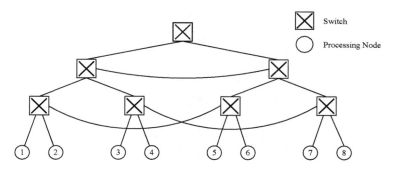

FIGURE 1.12

A hypertree interconnect with intermediate switches joint by links.

tree, while those that can use the symmetry of the n-cube can be assigned to processors in a way that efficiently uses the n-cube links. Figure 1.12 shows a hyper tree interconnect with eight processing nodes and seven switches.

The root has a node address 1. Addresses of left and right children are formed by appending a "0" and "1," respectively, to the address of the parent node (i.e., the children of node x are numbered $2x$ and $2x + 1$). As in half-ring or full-ring structures, additional links in a hyper tree are horizontal and connect nodes that lie on the same level of the tree. In particular, these links are chosen to be a set of n-cube connections, connecting nodes that differ by only one bit in their addresses [69].

A performance-defining property of an IN is the distance messages must travel. It is advantageous to make this parameter as small as possible because it not only reduces traveling time for messages, but also minimizes message density (i.e., used bandwidth) on the links. For a hyper tree, the longest distance is 1/2 and 1/3 of the distance in a simple binary tree. Absolute worst-case distances between any two nodes in a hyper tree can be identified on the observation that no path goes higher than the middle level of the tree.

The regular structure of this IN allows the adoption of simple routing algorithms, which require no detailed knowledge of the network. However with a relatively small additional overhead, a robust routing algorithm can be built so that messages arrive at the proper node, even for largely unbalanced trees or trees in the presence of nodes or link failures. All nodes have a fixed number of ports regardless of the size of the network, which ensures extensibility of the system and the second ensures graceful degradation in the presence of communication hardware failures [69].

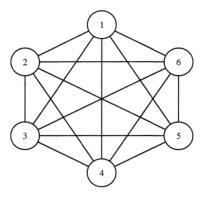

FIGURE 1.13
A fully connected network with six nodes.

1.2.6   Fully Connected Network

The fully connected topology for the parallel interconnect can be viewed as a complete graph, where every node is connected to every other node, as Figure 1.13 shows. A network with n processing nodes has $n(n-1)/2$ edges. This network is also referred to as a full mesh interconnect.

A direct link from one node to every other node results in a diameter of 1. The degree of a node is n (the number of nodes), and it increases as the network size grows. The degree dampens the scalability of this network as the number of links is large. The bisection width is n as there is a link from one to every other node. The advantage of this network is the diameter and the simplicity to route messages from one node to another [183].

1.3   Indirect Networks

Indirect interconnection networks are built using switching elements to provide connectivity to individual processing nodes. The network can have a single interconnection stage. However, cascading successive stages of switches may offer more flexibility and are more popular. Such networks are also referred to as multistage interconnects. Depending on the topology, a switch either connects to additional switches or to processing nodes.

An indirect interconnect may provide exactly a single path (Banyan, Delta) or multiple paths (Clos, Beneš) from source to destination. The source and the destination of a message in an IN are generally the processing nodes connected at the extremities of the network. The performance of the network is derived from the topology, switching and routing mechanisms, and the char-

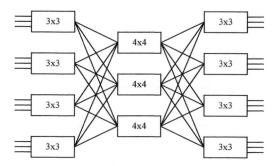

FIGURE 1.14
An indirect network: the Clos.

acteristics of that network. Like the direct interconnect, indirect networks can be differentiated based on key properties of the network (i.e., the topology), switching, and routing characteristics. Using graphs to model the INs onto a plane helps us derive some of the topological characteristics of each interconnect network. The network formed as a result of the connectivity between the multiple stages of the switching elements is also sometimes referred to as dynamic interconnect because the resulting interconnect pattern between stages can change with time. This change can be controlled by a centralized processor or through distributed algorithms. A classic example of an indirect network is the Clos network [42], although this network is now used as an IN for computer systems, the Clos network was originally proposed by Charles Clos in 1952 to interconnect multistage circuit switched telephony systems. It was designed to be a nonblocking interconnect (under some conditions) and used fewer crosspoints as compared to a crossbar switch. Figure 1.14 shows an example of an 8x8 Clos interconnect. The labels in the small switches in the figure show the size of the switches, in the number of inputs times the number of outputs.

1.3.1 Single-Stage Interconnection Networks

1.3.1.1 Crossbar Network

A single-stage network allows a set of nodes (processor or memory) to connect to each other using a single switching element. An example of a single-stage IN is the crossbar network, which allows multiple nodes to communicate simultaneously and without contending for the switching infrastructure once a switching resource is reserved. Figure 1.15 shows a crossbar interconnect with N inputs and M outputs. New connections can be established at any time under the constraint that the requested sets of input and output ports are free.

A crossbar network has N inputs and M outputs. It allows up to a mini-

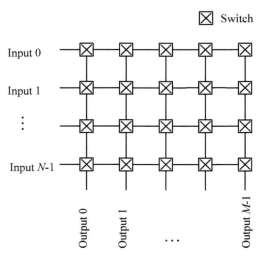

FIGURE 1.15
An *N*x*M* crossbar interconnect.

mum of (N,M) contention-free interconnections. The *N*x*M* crossbar network has NM switching points (or crosspoints). Note that this product gives a large number of crosspoints for large N and M, making the crossbar an expensive IN for a large number of inputs and outputs. The cost of the crossbar is $O(NM)$.

Crossbar networks have been traditionally used in small-scale shared-memory multiprocessors, where multiple processors are allowed to access memories simultaneously as long as each processor reads from or writes to a different memory. When two or more processors contend for the same memory module, arbitration allows a processor to proceed while the others wait. The arbiter in a crossbar is distributed among all the switch points connected to the same output, and the arbitration scheme can be less complex than one for a shared bus [57].

Crossbar networks are used in the design of high-performance small-scale multiprocessors, in the design of routers for direct networks, and as basic components in the design of large-scale indirect networks. A crossbar network employs a distributed control, where each switch point may have one of four states, as Figure 1.16 shows. In Figure 1.16 (a), the input from the row containing the switch point has been granted access to the corresponding output, while the inputs from upper rows requesting the same output are blocked. In Figure 1.16 (b), an input from an upper row has been granted access to the output. The input from the row containing the switch point does not request that output and can be propagated to other switches. In Figure 1.16 (c), an input from an upper row has also been granted access to the output. However, the input from the row containing the switch point also requests that output

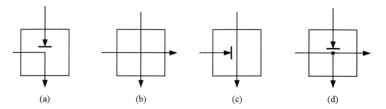

(a)   (b)   (c)   (d)

FIGURE 1.16
Switch states for a crossbar interconnect.

and is blocked. The configuration in Figure 1.16 (d) is only required if the crossbar has to support multicasting (i.e., one-to-many communication) [57].

1.3.2 Multistage Interconnect Networks

Multistage interconnection networks (MINs) provide a set of interconnects between processing nodes and shared-memory modules in SIMD architectures or just processors in the case of an MIMD architecture by cascading multiple switching stages, joined by interconnecting links. One end of the set of these cascaded switches forms the input stage of the network, which is then linked through at least one, but often more, intermediate stages of connecting switches to the output stage of the interconnect network. Cascading multiple switch stages helps reduce the permutation complexity of connecting multiple stages in the network [48]. The resulting connection patterns between each stage in the network are responsible for defining the routing capabilities of the whole network. The primary motivation for using multistage interconnects is to minimize the cost-to-performance ratio in comparison to single-stage crossbar networks where the cost incurred is high because the number of crosspoint switches is a quadratic function of its inputs (and outputs). However, it is relevant to note that the internal construction of each switch in a multistage interconnect forms a crossbar and the connection scheme between neighboring stages together with the number of stages defines the properties for the whole network. Typically multistage interconnect networks are used to connect asynchronous processing nodes, which rely on routing algorithms to communicate within the network. Multistage interconnect networks may also be used in packet networks to improve the total achievable bandwidth from input to output [48].

1.3.2.1 General Form of a Multistage Interconnect

A general form of the multistage interconnect with N inputs and M outputs uses G stages, where G_0 to G_{g-1}, and C_{g-1} gives the connection pattern in-between stages. Figure 1.17 shows the general form of the multistage in-

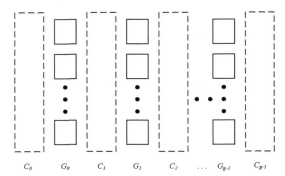

C_0 G_0 C_1 G_1 C_2 ... G_{g-1} C_{g-1}

FIGURE 1.17
A general form of the multistage interconnect.

terconnection network. The set of interconnect links C_0, where the processing nodes (or memory) can be connected in an alternating pattern of interleaved stages, ends with the link pattern C_g. This form can be used to describe other interconnection networks. However, there are variations in the number of inputs and outputs for each switching stage, number of intermediate switches, and the pattern of different interconnects, which result in networks with different characteristics. A set of processing nodes or shared memory (depending on the system architecture) would be connected to the outputs of the last stage of the IN.

1.3.3   Unidirectional MINs

Multistage networks can be broadly categorized as unidirectional or bidirectional networks. A unidirectional MIN consists of a series of multiple switching stages (with at least two), where packets always move from the input links, at the left side, to exit links on the right side. To allow bidirectional communication between nodes each input/output pair is connected to a single node, usually by a short link to the ingress and a long link to the egress. Unidirectional MINs are designed in such a way that it is possible to move from any point of entry to an exit in a single pass through the network.

A large number of connectivity permutations (i.e., different configurations of interconnecting switches at neighboring stages) have been proposed for unidirectional MINs. These include the shuffle-exchange, the Butterfly, and Omega networks, as examples. These topologies are isomorphic, or topologically equivalent. This means that any one topology can be converted to any of the other topologies simply by moving the switching elements and the end points up and down within their stages. Aside from these topologies, there are other unidirectional MINs, such as the Beneš network, that provide more than one path between each pair of nodes.

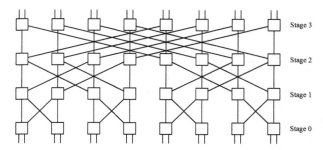

FIGURE 1.18
A four-stage butterfly network.

Each of the existing topologies presents a trade-off between implementation complexity, blocking vs. nonblocking, the ability to route from a single source to multiple destinations in one pass through the topology, partition ability, fault tolerance, and the number of switching stages required to realize these properties. An important characteristic of unidirectional multistage networks is that the path length to any destination is constant. This is an important characteristic for scheduling packets, as it helps to make some network features predictable.

1.3.3.1 Butterfly (k-ary n-fly) Network

The butterfly or k-ary n-fly network is one of the most commonly mentioned topologies in multistage networks. Here, k and n represent the number of ports in each direction of every switch and the number of switching stages, respectively. Figure 1.18 shows this interconnection pattern.

This figure shows that an outgoing switch port is identified by an n-digit number (tuple) where each digit x may have the values from 0 to $k-1$ (x_{n-1}, $x_{n-2}, \ldots, x_0$). In this notation, the sequence $x_{n-1}, x_{n-2}, \ldots, x_1$ corresponds to the switch number and x_0 correspond to the switch port number. The connection from port $a_i - 1$ at stage $i - 1$ (enumerated from 0 on the left) to a switch $b_i = (b_{n-1}, b_{n-2}, \ldots, b_1)$ at stage i is given by exchanging the first digit x_0 with the ith digit x_i. For example, port 000, which is port 0 on switch 00 at stage 0, is connected to switch 00 at stage 1. Port 001, which is port 1 on switch 00 at stage 0, is connected to switch 01 at stage 1, and so on. A butterfly network of any size may be constructed in this fashion.

1.3.3.2 Omega Network

The omega network represents another well-known type of MIN and was proposed for SIMD machines. It is based on the perfect-shuffle interconnection pattern [166]. A size N omega network consists of n ($n = \log_2 N$ single stage) shuffle-exchange networks (SENs). Each stage consists of a column of $N/2$

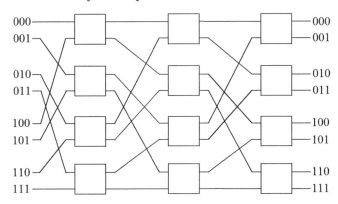

FIGURE 1.19
An 8x8 omega network.

two-input switching elements whose input is a shuffled connection. Figure 1.19 illustrates the case of an $N = 8$ omega network. As the figure shows, the inputs to each stage follow a shuffle interconnection pattern. Note that the connections are identical to those used in the 8x8 SEN in Figure 1.20.

1.3.4   Bidirectional MINs

Bidirectional networks are usually derived from unidirectional MINs by folding. That is, end nodes of a MIN are connected to only one side of the topology. To reach any other end node, packets are forwarded to the right side of the network until they reach a switch that is also reachable by the destination node, at which point the direction reverses and the packet travels to the left, towards its destination. Therefore the resulting path length from a specific source to different destinations may be different in a bidirectional interconnect network. Here, the destinations can be grouped by locality with respect to a specific source, yielding a hierarchical representation of the distance of the destinations.

1.3.4.1   Fat-Tree Network

A fat-tree is a routing network based on a complete binary tree, as Figure 1.9 shows. A set P of n processors is located at the leaves of the fat-tree. Figure 1.21 shows an 8-node fat-tree network. Each edge of the underlying tree corresponds to two channels of the fat-tree: one from parent to child, the other from child to parent. Each channel consists of a bundle of wires, and the number of wires in a channel builds its total capacity. The capacities of the channels of a fat-tree network grow exponentially as we go up the tree from the leaves. Initially, the capacities double from one level to the next, but at levels closer to $3 \log(n/w)$ of the root, where w is the root node capacity, the

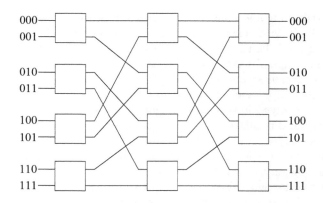

FIGURE 1.20
An 8x8 shuffle-exchange network.

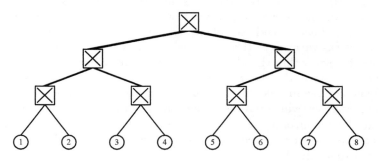

FIGURE 1.21
An 8-node fat-tree interconnect.

channel capacities grow at the rate of $3\sqrt{4}$ [102]. For the root node, the channel leaving the root of the tree corresponds to an interface with the external world. Each (intermediate) node of the fat-tree contains circuits that switch messages between incoming and outgoing channels. A characteristic of the fat-tree is that it is said to be universal for the amount of interconnection hardware it requires, in the sense that any other routing network of the same volume can be efficiently simulated. Here, *efficiency* is defined as *at most poly-logarithmic slowdown* in the network [102].

1.3.5 Design Constraints

The choice of an optimal network for a multiprocessor system must be sensitive to the system parameters and the constraints. System parameters include, among others, message size and the degree of communication locality; communication locality in parallel programs depends on the algorithms used as well as on the partitioning and placement of data and processes. Technological constraints also limit wire density and impose physical limitations on wire signal propagation speeds. Thus, the topology of an interconnect network is governed by the system design objectives and the limitations imposed by technology for a specific application scenario.

All the above topologies discussed so far are characterized by having their network links arranged in several orthogonal dimensions in a regular fashion. The symmetry and regularity of these topologies simplify network implementation (i.e., packaging) and packet routing as the movement of a packet along a given network dimension does not modify the number of remaining hops in any other dimension toward its destination.

Direct networks, with the exception of the fully connected network, may introduce blocking among packets that concurrently request the same path, or part of it. The hop count in direct networks can likewise be reduced by increasing the number of topological dimensions via an increased switch degree. It helps to maximize the number of dimensions for a system of a certain size and switch cost but the electronic hardware such as the integrated circuit chips, printed circuit boards, and backplanes are for the most part bidimensional. Direct networks with up to three dimensions can be implemented using relatively short links within the 3D space, independent of the system size. Links in higher-dimensioned networks would require increasingly longer wires or fiber. This increase in link length with system size is also indicative of MINs, including fat-trees, which require either long links within all the stages or increasingly longer links as more stages are added. Besides link length, other constraints as governed by the topological characteristics may limit the degree to which a topology can scale [79].

The available pin-out is a local restriction on the bandwidth of a chip, printed circuit board, or the backplane connector. Specifically for a direct network that integrates processor cores and switches on a single chip or multichip module, pin bandwidth is used both for interfacing with main memory and for implementing node links. In this case, limited pin count could reduce the number of switch ports or bit lines per link. In an indirect network, switches are implemented separate from processor cores, allowing most of the pins to be dedicated to communication bandwidth. However, as switches are grouped onto boards, the aggregate of all input-output links of the switch fabric on a board for a given topology must not exceed the board connector pin-outs [79].

The bisection bandwidth is a more global restriction that defines the interconnect density and bandwidth that can be achieved by a given packaging technology. Interconnect density and clock frequency are related to each other: when wires are packed closer together, crosstalk and parasitic capacitance increase, which usually impose a lower clock frequency. For example, the availability and spacing of metal layers limit wire density and frequency of on-chip networks, and copper track density limits wire density and frequency on a printed circuit board. To be implementable, the topology of a network must not exceed the available bisection bandwidth of the implementation technology. Most networks implemented to date are constrained more so by pin-out limitations rather than bisection bandwidth, particularly with the recent move to blade-based systems. Nevertheless, bisection bandwidth largely affects performance. For a given topology, bisection bandwidth is calculated by dividing the network into two roughly equal parts, each with half the nodes and summing the bandwidth of the links crossing the imaginary dividing line. For nonsymmetric topologies, bisection bandwidth is the smallest of all pairs of equal-sized divisions of the network [79].

1.4 Further Reading

There is a large number of works about INs for multiprocessors in the literature. A large collection of INs may be found in [47, 57].

1.5 Exercises

1. Draw a 2D mesh network with 36 nodes.
2. What is the diameter of the network in Exercise 1?
3. What is the switch degree of a mesh network?
4. What is the switch degree of a 2D torus network?
5. What is the bisection of a torus 2D network?
6. What is a nonblocking network?
7. What is the number of nodes in a honeycomb network of size 4?
8. What is the diameter of a binary tree with height 4?
9. Mention a drawback of a multistage interconnection network.
10. Mention a disadvantage of a fully connected network.
11. Draw a 8x8 butterfly network.
12. Mention a disadvantage of omega networks.

2

Routing

CONTENTS

Routing is basically finding a path to reach the destination node from the source node in a particular topology. As an analogy, routing is to select a route to reach the destination point departing from a starting point using a given road map. This chapter presents several routing schemes used in interconnected systems and networks.

2.1 Introduction

Routing algorithms typically aim to yield the shortest path(s) in terms of hop count, latency, or any other metric between the source and the destination nodes. Many efficient routing algorithms aim to balance the traffic load on the channels/links of a network even under traffic with nonuniform distributions.

An effective load-balancing mechanism can make the network throughput to approach the ideal one.

Another aspect of routing is fault tolerance. If a routing algorithm can adapt to node/link failures, the whole interconnection system can keep functioning in the presence of a failure, but possibly with a slight performance degradation. In addition, the possibility of deadlock/livelock in routing algorithms should also be considered. If a routing algorithm is not carefully designed, deadlock/livelock may occur.

Routing algorithms can be categorized into three classes according to how the set of possible paths is selected by the routing algorithm: (a) deterministic (static), (b) oblivious, and (c) adaptive routing. These three categories of routing algorithms are described below.

2.2   Deterministic (Static) Routing

Deterministic routing always chooses the same path between the source-destination pair (assuming a one-to-one connectivity), even if there are multiple optional paths. Such a routing approach can cause congestion if the traffic is not homogeneous and several messages from different sources are routed through the same link. Static routing algorithms are ineffective at load balancing because they simply neglect any path diversity of the topology. However, these schemes are common in practice because they can be easily implemented and made deadlock-free. A positive feature of deterministic routing is that packets transmitted between each source-destination pair are delivered in order. Popular deterministic routing algorithms are destination-tag routing in butterfly networks [99] and dimension-order routing for cube networks [167]. These are described as follows.

2.2.1   Destination-Tag Routing in Butterfly Networks

Destination-tag routing uses the destination node's address to determine the path taken by a packet(s) to reach its destination [99]. In a k-ary n-fly network, the destination address is interpreted as an n-digit radix-k number and each digit of this number is used to select the output port of the switch at each step. Here, k denotes the number of input and output ports for each switch and n is the number of stages in a butterfly network. Figure 2.1 delineates a 2-ary 3-fly network. The bold lines show the path taken to reach the destination node $3 = 011_2$ from source node $6 = 110_2$ using destination-tag routing. In this figure, the binary representation of destination node's address, 011, is used, and the upper and lower output ports of the switches at each stage are followed consecutively. In general, the radix-k representation of the destination address is used to route packets in any butterfly network with different k values.

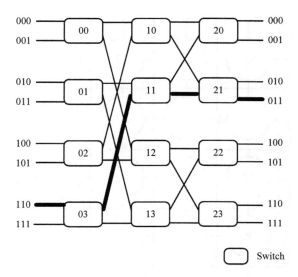

FIGURE 2.1
2-ary 3-fly network and the selected path between the nodes 110 and 011 by
the destination-tag routing.

2.2.2 Dimension-Order Routing in Cube Networks

Dimension-order routing in cube networks is analogous to destination-tag
routing in butterfly networks [167]. Dimension-order routing in cube networks
is also called *e-cube routing* and it is designed for k-ary n-cube networks (tori
and meshes), where k is the number of nodes in each row (or column) and n
is the dimension of the cube network. Dimension-order routing first computes
the preferred directions that are to be followed by packets to find a shortest
path to the destination node. For instance, in a 2-dimension (2D) cube network, the preferred directions would be indicated by a (x, y) pair, where x and
y can be -1, $+1$, or 0 to indicate the negative, positive, or either direction in
each dimension, respectively. Figure 2.2 shows a 4-ary 2-cube network and the
path taken by dimension-order routing to reach the destination node, $d = 12$,
from the source node, $s = 20$, with bold lines. To find the preferred directions
for each digit i of the address of the source and destination nodes, a relative
address, Δ_i, is calculated:

$$m_i = (d_i - s_i) \mod k \tag{2.1}$$

$$\Delta_i = m_i - \begin{cases} 0 & \text{if } m_i \leq k/2 \\ k & \text{otherwise} \end{cases} \tag{2.2}$$

After the calculation of the relative address, the preferred directions, D_i,
for each dimension can be calculated as follows:

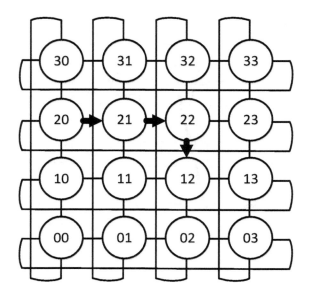

FIGURE 2.2
4-ary 2-cube network and the selected path between the nodes 20 and 12 by
the dimension-order routing.

$$D_i = \begin{cases} 0 & \text{if} \quad |\Delta_i| = k/2 \\ \text{sign}(\Delta_i) & \text{otherwise} \end{cases} \tag{2.3}$$

If Equations 2.1 and 2.2 are used to find one of the available shortest paths
from $s = 20$ to $d = 12$ in Figure 2.2, $m = (3,2)$ and $\Delta = (-1,2)$. By using
Δ, the preferred directions are calculated as $D = (-1,0)$, which means either
positive (i.e., clockwise) or negative (i.e., counterclockwise) direction in the
x-dimension and the negative direction in y-dimension must be followed.

2.3   Oblivious Routing

In oblivious routing the path(s) to be taken is completely determined by the
source and the destination addresses without considering the network's current
state (e.g., current traffic load on the links). Therefore, oblivious routing can
be considered as a superset of the set of the deterministic routing algorithms.
For instance, a random routing algorithm, which uniformly distributes traffic among all possible paths, can be classified under the category of oblivious
routing, as random routing does not consider the present state of the network.

Oblivious routing can be divided into three subcategories: randomized routing, minimal oblivious routing, and load-balanced oblivious routing. Randomized routing is based on Valiant's algorithm and it balances the load well. However the locality of the traffic may be lost. Minimal oblivious routing preserves the locality of the traffic and generally improves the throughput of the network. On the other hand, minimal routing may not balance the load well, and thus, minimal oblivious routing yields worse throughput than Valiant's randomized routing in torus networks. Load-balanced oblivious routing produces a mid-point between the randomized and the minimal oblivious routing algorithms in terms of the load balancing and preserving locality capabilities.

2.3.1 Valiant's Randomized Routing Algorithm

Valiant's randomized routing algorithm simply selects a random intermediate node between the source and the destination where all the packets from the source node are first sent [174]. Then, the packets are routed from the intermediate node to their destination. By randomly selecting an intermediate node, Valiant's randomized routing can balance all types of traffic on every connected topology. For both stages (i.e., *source → intermediate node* and *intermediate node → destination*) of the Valiant's randomized routing, any arbitrary routing algorithm can be employed. Nevertheless, a routing algorithm that balances the load under uniform traffic yields the best result in terms of load balancing for both stages. Therefore, the dimension-order routing for torus and mesh networks and the destination-tag routing for butterfly networks fit well in both stages.

2.3.1.1 Valiant's Algorithm on Torus Topologies

Although the locality of the traffic is lost and the two-stage routing introduces an overhead, Valiant's randomized routing algorithm yields good worst-case results in k-ary n-cube networks. For instance, the nearest-neighbor traffic, which normally requires only one hop to reach the destination when a minimal routing algorithm is used, requires $nk/2$ hops when Valiant's randomized routing is employed. The two-stage nature of Valiant's randomized routing may introduce an extra delay and decrease the throughput compared to minimal routing algorithms. Figure 2.3 shows a routing example where Valiant's randomized algorithm is employed between the source node, 20, and the destination node, 12. In the example Figure 2.3 shows, it is assumed that the randomly selected intermediate node is 01 and the employed routing algorithms from node 20 to 01 and from 01 to 12 are both dimension-order routing, which is introduced in Section 2.2.2. Notice that if the minimal routing were used it would require three hops to reach the destination node, 12, as Figure 2.3 shows. However, Valiant's randomized routing algorithm requires five hops and, therefore, these additional hops increase the delay for each packet.

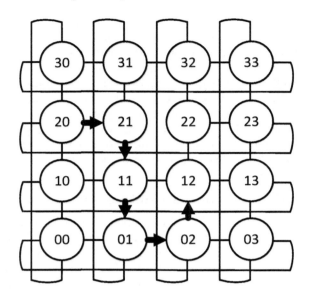

FIGURE 2.3
Selected path between the source 20 and the destination node 12 by Valiant's randomized routing algorithm in a 4-ary 2-cube network, if randomly selected intermediate node is 01.

2.3.2   Minimal Oblivious Routing

Minimal oblivious routing [120] aims to balance the load and preserves the locality of the traffic by restricting the routes to be minimal (i.e., the shortest paths). Minimal oblivious routing balances the load well for hierarchical topologies, such as fat-tree, while it preserves the locality of the traffic. In this section, we will explore how minimal oblivious routing algorithm works on a folded-Clos (fat-tree) and torus networks.

2.3.2.1   Minimal Oblivious Routing on a Folded-Clos Network (Fat-Tree)

Figure 2.4 shows a folded-Clos network with 16 nodes (terminal). This type of network is also called folded Beneš network with concentration, or fat-tree, where each pair of terminals are connected to a concentrator. In Figure 2.4, each concentrator has 2:1 concentration ratio and has a corresponding address that matches with its connected terminals. For instance, the first concentrator, on top in Figure 2.4, has an address 000X to indicate that it is connected to terminal 0000 and terminal 0001, where the last digit X denotes a don't care value, which can be equal to either 0 or 1. All the concentrators are then connected to an 8-port radix-2 folded-Clos (Beneš) network, as Figure

2.4 shows. Each switch of the Beneš network has an address pattern that represents the terminals reachable from the left. In other words, each switch covers all the address space belonging to terminals connected to the switch from the left. For instance, a switch labeled with 1XXX covers all the terminals within the address space in [1000, 1111], as the figure shows.

Minimal oblivious routing on a folded-Clos network first finds the nearest common ancestors of the source and the destination nodes [157]. Second, one of those ancestors is randomly chosen and the packet is routed to the chosen ancestor. Finally, the packet is sent to its destination by the selected ancestor via a unique shortest path. In Figure 2.4, two possible shortest paths are marked with bold solid and bold dashed lines between the source node, 8 (represented as 1000 in binary form), and the destination node, 14 (represented as 1110 in binary form), if the minimal oblivious routing is employed. Two nearest common ancestors of nodes 8 and 14 are the switches labeled with 1XXX A and 1XXX B. If the minimal oblivious routing randomly selects switch 1XXX A as the common ancestor, the shortest path follows the solid bold lines; otherwise (i.e., selecting switch 1XXX B), the shortest path follows the dashed bold lines in the figure. Notice that in either case, the selected path would be the shortest and the total number of traversed hops remains the same, which is equal to 6.

Minimal oblivious routing can be carried out incrementally by randomly selecting the upper right or lower right output port of a switch until the packet reaches to a common ancestor for both the source and the destination node. Then, the packet is routed using a partial destination-tag routing from that ancestor switch to its destination. The partial destination-tag routing mentioned here uses the binary digits of the destination node's address which corresponds to the don't-care digits of the common ancestor switch. For the incremental version of the minimal oblivious routing in Figure 2.4, when the packet reaches switch 1XXX A (or 1XXX B), minimal oblivious routing uses the last three digits of the destination address, 110, to route the packet to its destination. Therefore, the lower left port of the switch 1XXX A (or 1XXX B), lower left port of the switch 11XX, and the upper left port of the switch 111X are sequentially selected to reach the destination.

It is important to note that by randomly selecting a nearest common ancestor switch among the possible ancestors, the load between them is balanced. Moreover, if any other route is followed rather than the shortest path yielded by the minimal oblivious routing, load balancing is not improved and bandwidth consumption increases.

2.3.2.2   Minimal Oblivious Routing on a Torus

Minimal oblivious routing simply restricts the selection of the intermediate node to a region called minimal quadrant, where the diagonal corners of that region are the source and the destination nodes. Figure 2.5 shows two possible minimal quadrants as shaded rectangles, if the source and the destination

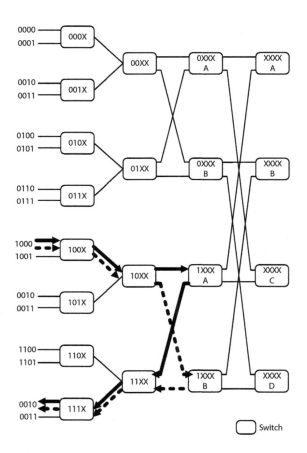

FIGURE 2.4
Folded-Clos network (fat-tree) with 16 nodes and the possible shortest paths between node 8 (1000 in binary) and 14 (1110 in binary) yielded by the minimal oblivious routing.

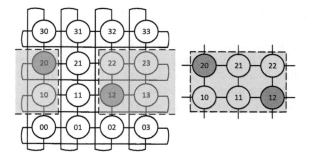

FIGURE 2.5
Possible minimal quadrants for minimal oblivious routing in a 4-ary 2-cube network, if the source and the destination nodes are selected as 12 and 20, respectively.

nodes are selected as 20 and 12, respectively. Because both minimal quadrants yield the shortest path between the source and the destination, either of them can be chosen for the selection of the intermediate node. The size of the minimal quadrant is determined by the magnitude of the relative address Δ. Moreover, the preferred direction vector defines the position of the minimal quadrant considering the source node. In Figure 2.5, Δ is calculated as $(-1,2)$, and therefore x and y dimensions of both possible minimal quadrants span two and one hop, respectively. The minimal quadrants in this figure are located either to the left or to the right of the source node in the x-dimension ($D_0 = 0$) and they are both below the source node in the y-dimension ($D_1 = -1$). Once the intermediate node is randomly chosen from the minimal quadrant, packets are routed from the source to the intermediate node, and from the intermediate node to the destination using dimension-order routing. Note that the source and the destination nodes can be selected as intermediate nodes.

The load-balancing ability of minimal oblivious routing in torus networks depends on two aspects: the selected intermediate node and the first taken direction (i.e., x-first or y-first routing) to route the packets in dimension-order routing for both routing stages. Therefore, randomly selecting x-first and y-first routing may help to balance the load in addition to the random selection of the intermediate node. On the other hand, minimal oblivious routing on torus networks preserves locality at the expense of achieving the worst-case throughput for special types of traffic patterns, such as tornado traffic [172].

2.3.2.3   Load-Balanced Oblivious Routing

Load-balanced oblivious routing is a trade-off between the locality of the traffic and the load-balancing ability of the employed routing algorithm. Therefore, it can be placed as an approach between Valiant's randomized routing (completely randomized traffic) and minimal oblivious routing. Load-balanced oblivious routing chooses the quadrant to route by weighting the choice of quadrants considering the distance between the source and the destination. For each dimension the short distance is selected with a higher probability than the longer distance. However, in this case, there is a possibility for non-minimal quadrants to be selected in addition to minimal quadrants. Although the selection of non-minimal quadrants improves the load balance and outperforms the Valiants algorithm on local traffic, its worst-case throughput is lower than that of the Valiants algorithm.

2.4   Adaptive Routing

Adaptive routing [7, 35, 46, 108, 121] uses the information about the network state, such as queue occupancies and the traffic load on the links, to route the packets. Despite the deterministic and oblivious routing algorithms, adaptive routing is tightly coupled with flow control mechanisms because adaptive routing depends on the network's state.

It is expected that adaptive routing outperforms deterministic and oblivious routing algorithms. However, most practical adaptive routing algorithms are based on local information about the network's state. Using the local network-state information might lead to global load imbalance and poor worst-case performance compared to some deterministic and oblivious routing algorithms due to the local decisions to route the packets [20]. For instance, suppose that adaptive routing is employed to route a packet from the source node, 00, to the destination node, 22, in a 4-ary 2-cube network and there is one slightly congested link and two highly congested links as Figure 2.6(a) shows. Figure 2.6(b) shows the path chosen by adaptive routing after considering the local congestion information. The local routing decision at node 10 (Figure 2.6) forces the packet to be routed through the highly congested link between the nodes 12 and 22. This selection results in having the packet experiencing a long queueing delay. In addition to nonoptimal path selection, reaction to congestion may be delayed because the queue at the decision point must exceed a threshold or become saturated before reaction to congestion takes place. Moreover, the dissipation of the congestion messages may require to traverse more than one hop until it reaches the source node, and hence, the traveling time of these messages might introduce additional delays for reacting to congestion.

Usually routers that employ adaptive routing use queue occupancies to estimate the congestion of local links. In this case, it may not be possible for a switch/node to sense the congestion occurring on one of the links that are not directly connected to the node that senses the congestion first (i.e., sensing node). Since only the queue occupancy of the sensing node may not reflect the emerging/ongoing congestion on other link(s), it may presume that

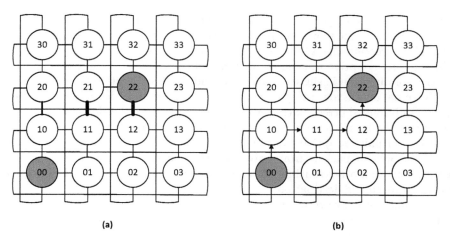

(a)   (b)

FIGURE 2.6
(a) A slightly congested link between nodes 10 and 20, and two highly congested links between the pairs 11, 21 and 12, 22 (the extent of the congestion is represented by the thickness of the lines). (b) Adaptive routing decides the route denoted by arrows between the source node, 00, and the destination node, 22, considering the local congestion information.

there is no congestion on the way and send packets through the congested link. On the other hand, *backpressure* is a mechanism used to disseminate the congestion signals back to the source node. When the queue occupancy of any node becomes filled or exceeds a threshold, the node sends a stopping signal to the one-hop-away ancestor switch/node in the direction opposite to the traffic flow. This congestion signal propagates back to the source node and all the switches along the way infer that there is congestion on the path in the direction of traffic flow. The delay to react to congestion increases if the buffers of the switches along the way are large to hold a large number of packets.

2.4.1 Minimal Adaptive Routing

Minimal adaptive routing uses information about the network state and makes the route selection among the possible shortest paths between the source and the destination node [103]. The decision mechanism of minimal adaptive routing runs at each hop and each node along the path first produces an output array, called *productive* vector, which indicates the links needed to get the packet closer to its destination. Network state information is then used to select the best possible link from the productive vector according to different network state metrics, such as queue occupancy. In Figure 2.6, minimal adaptive routing yields a productive vector of (1,0,1,0) to indicate the $+x$ and $+y$ directions that can move the packet closer to its destination among all the available directions, $(+x, -x, +y, -y)$. Then, the $+x$ direction is selected by the minimal adaptive routing algorithm and the packet is routed to node 11, as the $+y$ direction is slightly congested and the $+x$ direction is idle.

Minimal adaptive routing may not avoid congestion if the minimal routing yields only one shortest path between the source and the destination node. In other words, if there is no minimal path diversity between the source-destination pair, adaptive routing may be forced to select that path and it does not prevent packets from traversing the congested link(s). Nevertheless, nonminimal adaptive routing can avoid congested link(s) by selecting longer paths to reach the destination.

Minimal adaptive routing can be applied to all type of topologies in addition to torus topology. For instance, in a folded-Clos network, an adaptive routing mechanism can be used as a two-phase routing. First, the packet is routed adaptively from left to right until a common ancestor of the source and the destination node is reached. The packet is then deterministically routed from right to left until it reaches its destination.

2.4.2   Fully Adaptive Routing

Fully adaptive routing allows longer paths (unproductive ones) to be selected in addition to the shortest paths (productive ones) to avoid congested or failed links [62]. In this case, packets may travel longer distances but the performance can be higher than that of local adaptive routing, since the packets do not have to traverse the congested links. Fully adaptive routing prioritizes the productive links in the absence of congestion, but it can select the unproductive links when the congestion exists on all the productive links. A possible fully adaptive algorithm can be as follows: First, a productive output port is selected to forward the packet if any of those ports has a queue occupancy smaller than a predefined threshold. Otherwise, without considering it is a productive or unproductive output, the output port with the shortest queue length is used to forward the packet. To avoid delivering the packet to the node from which it just arrived, some algorithms do not allow to send packets following a U-turn path (i.e., backtracking). Figure 2.7 shows how fully adaptive routing can avoid a congested link by selecting a path longer than the shortest one using the path $00 \rightarrow 10 \rightarrow 11 \rightarrow 21 \rightarrow 20$, where the source node is 00 and the destination node is 20. Forwarding a packet using such a longer path is usually called *misrouting*.

A drawback of fully adaptive routing is that it may lead to a livelock, unless the preventive steps against livelock are taken. Livelock is the phenomenon where a packet never reaches its destination and travels indefinitely on the network. Figure 2.8 shows an example about how a packet may be trapped in an indefinite loop due to congested links along the possible productive and unproductive links. Fully adaptive algorithm may resolve the livelock problem

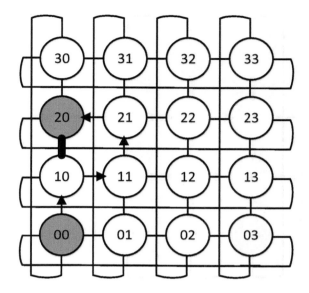

FIGURE 2.7
Full adaptive routing avoids the congested link by selecting a longer path, when the source node is 00 and the destination node is 20.

by ensuring progress over time. One possible solution is to limit the number of misrouting attempts when the misrouting threshold is exceeded. In this case, the packet is forced to be routed using minimal adaptive routing. Another possible solution is to allow to misroute the packet one hop for every $x > 1$ productive hops. A third possible solution is called *chaotic routing*, in which the router randomly grants only one packet among the contending packets for the same link. The rest of those contending packets lose the contention and they are misrouted to any available output port. This approach probabilistically provides livelock-free packet delivery. The probability of delivering the packet in K cycles approaches zero when K increases.

2.4.3 Load-Balanced Adaptive Routing

Adaptive routing algorithms usually do not balance the load across the available links along the destination node. Global load imbalance occurs because adaptive routing algorithms generally make decisions based on the local information as this is more practical and simpler to do. A possible approach to handle load imbalance is to first select a quadrant using the load-balanced oblivious routing (introduced in Section 2.3.2) and then use the adaptive routing within this quadrant. In such a hybrid approach, the selection of the quadrant using load-balanced oblivious routing aims to balance the global load, whereas

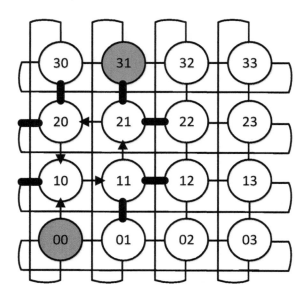

FIGURE 2.8
Livelock for fully adaptive routing, when the source node is 00 and the destination node is 31. Bold thick lines represent the congested links.

adaptive routing within the quadrant targets local load balancing. Moreover,
although this hybrid approach yields a good load balance, the agnostic nature
of oblivious routing may result in longer paths for some of the packets, and
hence, the performance of this hybrid routing may not produce better results
as compared to pure adaptive routing algorithm on local traffic patterns. It
additionally guarantees the packets always make progress to their destination,
so livelock is not an issue in load-balanced adaptive routing.

2.4.4   Search-Based Adaptive Routing

In this category, the problem of routing is approached as a search problem to
find the best possible path to destination by even backtracking (i.e., sending
the packet back to the node where it just arrived) or by first probing multiple
links before sending the packet and then transmitting the packet over the best
of these paths. However, because search-based routing algorithms are slow and
consume many resources, they are rarely used in practice. On the other hand,
they may be used to build routing tables off line.

2.5 Exercises

1. Find the switch IDs and their output ports (i.e., upper or lower) along the path to the destination node, $0 = 000_2$, from the source node, $7 = 111_2$, when the destination-tag routing is employed in a 2-ary 3-fly network given in Figure 2.1.

2. Calculate the relative address, $\Delta = (\Delta_1, \Delta_0)$, and the preferred directions, $D = (D_1, D_0)$, and calculate the path taken when the destination-tag (e-cube) routing is employed with the source node, $s = 31$, and the destination node, $d = 02$, in a 4-ary 2-cube network.

3. Find out how many hops a packet must traverse if Valiant's randomized routing algorithm is used with the source, destination and the random selected intermediate nodes are $s = 10$, $d = 13$, and $x = 21$, respectively. Also, give the path taken if x dimension on each stage of e-cube routing is traversed first (x-first routing).

4. Show all the possible minimal quadrants if the minimal oblivious routing is employed between the source node, $s = 20$, and the destination node, $d = 03$, in a 4-ary 2-cube network. Also provide all possible shortest paths between s and d, considering the both x-first and y-first routing.

5. Suppose that the path $11 \rightarrow 12 \rightarrow 13 \rightarrow 23$ is selected by the minimal adaptive routing to route a packet from the source node, 11, to the destination node, 23, in a 4-ary 2-cube network with four highly congested links as shown in Figure 2.9. Find the productive vectors at each node (excluding the destination node) along the given path above and explain why the path $11 \rightarrow 12 \rightarrow 13 \rightarrow 23$ is selected by minimal adaptive routing. Is there any other possible path(s) to route the packet using minimal adaptive routing?

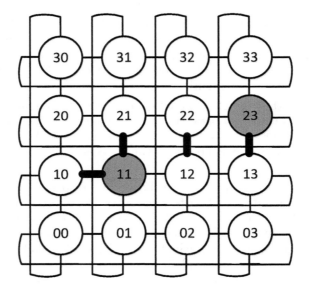

FIGURE 2.9

Minimal adaptive routing between the source node, 11, and the destination node, 23. Thick bold lines represent high congestion on those links.

Part II

Data Networks

3

Internet Protocol (IP) Address Lookup

CONTENTS

Finding out the egress port a packet traversing a network node, or router, requires performing the longest prefix matching between the destination of the packet to a prefix in the forwarding table. This process is called Internet Protocol (IP) address lookup, or just IP lookup. It requires to access this table to retrieve the next-hop information (NHI), which is the IP address or the router's egress port number where a packet is forwarded.

The forwarding table resides in the memory of the router. Every time a packet arrives in a router, memory storing the forwarding table is accessed to perform IP lookup. Unfortunately, memory is a component whose speed lags in comparison to the rates at which data can be transmitted through (optical) network links. This speed mismatch requires to perform lookup using the smallest number of memory accesses. In addition, to improve feasibility in building such schemes, the amount of required memory must be implementable. Furthermore, IP lookup mechanisms should be able to accommodate changes of the forwarding table caused by changes in routing tables, in an expeditious manner so as not to interrupt the lookup process.

3.1   Introduction

Classless Interdomain Routing (CIDR) [67] is an approach used to aggregate IP addresses as prefix. This aggregation of addresses has the objective of reducing the number of entries in a memory holding a forwarding table of an Internet router. CIDR uses a technique called supernetting to summarize a range of addresses represented as the combination of a prefix and prefix length. The prefix includes a range of addresses that are reachable by a given router port. Figure 3.1 shows an example of address aggregation at a router. In this figure, different networks, which are represented as ranges of addresses or prefixes, are reachable through different ports. Those addresses that share common most significant bits (MSBs) and are reachable through the same router port may be aggregated.

With CIDR, addresses can be aggregated at arbitrary levels, and IP lookup becomes a matching to the longest prefix. Specifically, a forwarding table consists of a set of IP routes. Each IP route is represented by a <prefix/prefix length> pair. The prefix length indicates the number of significant bits in the prefix, starting from the MSB. To perform IP lookup, the forwarding table is searched to obtain the longest matching prefix from among all the possible prefixes that match the destination IP address of an incoming packet. Then, the output port information for the longest matched prefix is retrieved and used for forwarding the packet. For example, a forwarding table may have three IP routes: <12.0.54.8/32>, <12.0.54.0/24>, and <12.0.0.0/16>. If an incoming packet has the destination IP address <12.0.54.2>, prefix <12.0.54.0/24> is the longest matched and its associated output port is retrieved and used for forwarding the packet.

Although CIDR resolves the aggregation of addresses to minimize the number of entries in a forwarding table (and in turn, in the memory of a router), the search for the matching prefix becomes more complex than performing exact matching (as used in label switching).

The complexity of address lookup is high because: 1) prefix lengths may be large and the adoption of IPv6 may increase these lengths even further, and 2) the number of prefixes is increasingly large for version 4 of the IP protocol, IPv4, and it is expected to grow with IPv6. In 1), the longest matching prefix is selected; so this selection requires a search for prefixes in all different lengths and to keep records of the longest prefix that last matched the destination address.

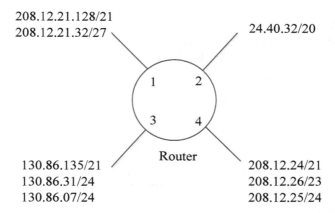

FIGURE 3.1
Forwarding information per port.

Prefixes have to be carefully stored and searched so as to minimize the number of times the memory storing the prefixes is accessed. The following example illustrates part of the search problem, whose complexity is affected by the relationship between the address of an entry and the entry content. Table 3.1 has eight entries (or rows) and each entry holds a number as content. Let's consider the contents of the table as a 20-number set, where each can take a value between 0 and 19. Note that the range of values an entry can have is a larger than the table size. Let's also consider that no entry has a duplicate. Let's assume that the entry contents are the set {1, 4, 7, 9, 10, 13, 18, and 19}, where the entries are stored in ascending order and are consecutively stored in the table, where 1 is stored in Entry 0, 4 in Entry 1, and so on. A simple strategy to find a desired number in the table would be to perform a linear search. For example, if we were to find number 15, we would go from location 0 to 6 (7 locations) and find out that 15 is not in the set. So the problem is to find a specific content in the smallest number of trials.

One could think of starting the search from different locations of the list (e.g., in the middle of the table) to reduce the number of accesses to the table. This problem exacerbates as the range of the content values or the number of entries grows. For example, in IP lookup, the contents could be found to be over half a million prefixes. This number is being reported by the time this book is written, but the number continues to increase [2].

Back in our example, we could sort the contents by selecting them based on the tens and the units of an entry, such that we could sort the numbers as a decimal tree where the first children would be from numbers from 0 to 1 (as root tree), and the children of those nodes would be numbers from 0 to 9. Note that if the set of numbers increases, so does the number of children in the root tree. We use here the terms tree or trie interchangeably in this chapter.

TABLE 3.1
Example of searching in a table with nonrelated set members.

Address	Content
0	1
1	4
2	7
3	9
4	10
5	13
6	18
7	19

TABLE 3.2
Example of prefixes in a forwarding table.

Prefix ID	Prefix
P1	*
P2	1*
P3	00*
P4	0001*
P5	1000*
P6	10001*
P7	11111*
P8	111011*
P9	100000*
P10	0001000*

Table 3.2 shows an example of a forwarding table. This table has 10 prefixes, with prefix lengths between 1 and 7 bits. Prefixes are represented by a string of 0s and 1s, marked at the end (right-most side) with a star mark (*). For instance, prefix 128/1 is represented as 1*.

3.2   Basic Binary Tree

In a simple form, prefixes can be stored in a binary tree representation (using 32 bits for IPv4 and 128 bits for IPv6).[1] This figure shows the binary tree of the forwarding table (Table 3.2), where the 0 child is at the left side of the root and the 1 child is at the right. he figure shows the position of the 10 prefixes in the binary three, where prefix **P1** has prefix length 0 (i.e., root of the tree), or it is positioned on level 0, and prefix **P10** is on level 7.

---

[1] Tree structures are covered in detail in [43].

The prefixes in the tree are stored, starting from the MSB to the least-significant bit (LSB). Furthermore, because we need to match any of the prefixes in the forwarding table, all entries are incorporated into the tree. Therefore, prefixes with common bits (bits with the same value at the same significant position) are shared. For example, prefixes 110* and 111* share the first two bits (11-) and the only different bit is their LSBs. So these two bits are represented as 11 linked to 0 for the first prefix and to 1 for the second. As this example shows, each position bit has only two possible combinations, or binary children. The default prefix, *, is the ancestor of all other prefixes (i.e., all address ranges overlap with the range represented by *). Therefore, * is the root of the three.

Figure 3.2 shows the binary tree representation of Table 3.2. Each time a packet arrives, the destination of the packet is compared bit to bit in the path from the root (*), or MSB, to a leaf, or LSB, of the tree. The matching prefix is the longest of those that matches the destination address of the packet.

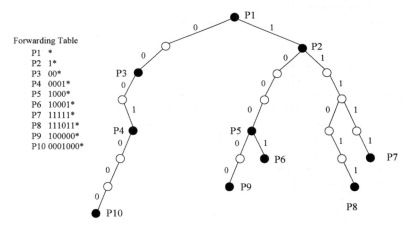

FIGURE 3.2
Binary tree of the prefixes in Table 3.2.

The binary tree of this table is stored in memory. Table 3.3 shows a simple implementation of the binary tree in Figure 3.2. In this representation, each memory location is assigned to a tree node (where the addresses of the table are the node numbers in the tree, not shown in the figure). A node may be represented with two children, the 0- and 1-child, as pointers. A node may be a parent node or prefix. If the node is a prefix a field for the prefix ID or NHI is also stored in that node. The memory address 0 stores the root of the tree. Each entry holds a flag indicating if the entry is a prefix or a non-prefix node, the Id of the node (or NHI), and two pointers to the address of the children of the node (children 0 and 1).

The table stores the 22 nodes of the tree. However, this table is built as a 32-entry (i.e., 2^5 memory block as the number of entry of memories come in power-of-two number of entries) table, the number of entries is equal to the number of tree nodes. The memory used to represent prefixes of any length must be of a feasible size (as fast memories, such as Static Random-Access Memory, SRAM, are small). SRAM is a suitable candidate memory technology for building lookup tables as it is desirable to perform IP lookup in the smallest possible packet inter-arrival time.

TABLE 3.3
An example of memory format storing the binary tree of Figure 3.2.

Memory address	Prefix flag	Prefix Id (NHI)	0-child	1-child
0	1	P1	1	2
1	0	-	3	-
2	1	P2	4	5
3	1	P3	6	-
4	0	-	7	-
5	0	-	-	8
6	0	-	-	9
7	0	-	10	-
8	0	-	11	12
9	1	P4	13	-
10	1	P5	14	15
11	0	-	-	16
12	0	-	-	17
13	0	-	18	-
14	0	-	19	-
15	1	P6	-	-
16	0	-	-	20
17	1	P7	-	-
18	0	-	21	-
19	1	P9	-	-
20	1	P8	-	-
21	1	P10	-	-

The matching process in a binary tree may require as many memory accesses as the longest stored prefix. This is up to 32 memory accesses for IPv4 and 128 for IPv6. The matching complexity is said to be in the order of $O(W)$, where W is the prefix length.

3.3   Disjoint Trie

The binary tree requires *remembering* the last matched prefix along the search for the longest prefix in the traverse from the root to the leaves. A register may be common to keep this temporary record. This additional memory requirement may be avoided if a *disjoint* trie is used instead. The disjoint trie has the property that each node has either two children or none.

Figure 3.3 shows the disjoint trie of the binary tree in Figure 3.2. As the figure shows, prefixes of a disjoint tree are at the leaves. Therefore, each time a packet arrives in a router, the tree is examined, started from the root (*) and continuing towards the matching leaves, and the visited leaf holds the longest matching prefix. This tree uses a number of nodes equal to or larger than that of a regular binary tree, but it simplifies the matching process by matching a single prefix.

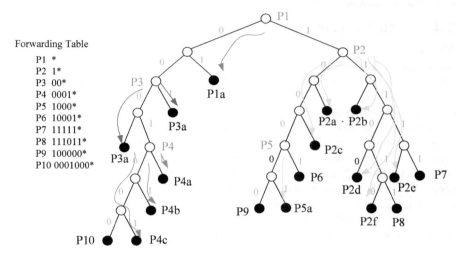

Forwarding Table

P1 *
P2 1*
P3 00*
P4 0001*
P5 1000*
P6 10001*
P7 11111*
P8 111011*
P9 100000*
P10 0001000*

FIGURE 3.3
Disjoint binary tree from prefixes in Figure 3.2.

3.3.1 Procedure to Build a Disjoint Trie

The disjoint trie is built by completing a binary tree. This process consist of making each node have two children or no children. This means that a node that has one child has to be completed with another child. Such an added node may be either a prefix or an ancestor node of a prefix (or a subtree with a prefix). In such cases, the new added child will hold the prefix of the closest ancestor prefix. This technique is also called *leaf pushing* [161].

For example, let's assume that the branch of a trie has the following two prefixes: **Px** 0010* and **Py** 00100* where 00100* is the only child (0 child) of 0010*. Therefore, 0010* must be completed by adding 1 child, or 00101*. Then 0010 is pushed from level four to level five of the trie as **Px** 00101*, and 0010 becomes an ancestor node (not a prefix) of these two prefixes that has two children.

The search complexity of matching on a disjoint tree is equivalent to that of a binary tree, this is, the worst-case scenario is 32 bits for IPv4 and 128 bits for IPv6. The prefix search complexity remains at $O(W)$. The amount of memory used for this trie is larger than the memory needed for a binary tree as the number of nodes increases as the tree is completed, but matching is achieved by reaching of the trie leaves as the trie no longer has internal prefixes (that could be matched).

3.4   Path-Compressed Trie

Although the comparison of bit by bit in a binary tree is faster than a linear search in IP lookup, the worst case scenario still takes time; 32 memory accesses for IPv4. The path-compressed trie [117], also called Patricia trie, improves the search for a prefix by comparing not one but many bits at a time. The number of compared bits may be variable in this trie; the number compared at once is equal to the number of compressed bits for a node. This technique is called path compression. In the Patricia trie, unbifurcated branches of a tree may be compressed. For example, if a forwarding table contains prefixes 1* and 111*, prefix 111* can be represented as a compressed prefix 11* with 1 bit compressed (indicated by the number of skipped bits, or *skip* field, or 1 bit) and by the value of the skipped bits, or *segment*="1" (the actual string of compressed bits). Figure 3.4 shows the Patricia trie of the binary tree in Figure 3.2. The compressed trie in this figure shows the same set of prefixes but in a smaller trie (i.e., a trie with a shorter height and a smaller number of nodes). However, the compressed bits and information indicating the compression must be stored for a node, indicating the skip and segment fields. In general, an entry of a Patricia trie uses additional space to store these two fields. That is, while the number of nodes is smaller (therefore, requiring a shorter memory), the memory required is wider than that of a binary trie. Reducing the tree height decreases the number of needed memory accesses.

In terms of performance, the Patricia trie performs lookup with a smaller number of memory accesses, on the average, than a binary trie. But in the worst-case scenario, the number of memory accesses may be equal to that of a binary trie. This worst-case scenario occurs when the path to the longest prefix in the table has bifurcations (two children) at each node. However, the possibility of this to occur may be small and may not occur in all possible paths.

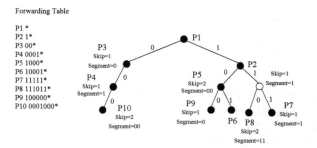

FIGURE 3.4
Example of a path-compressed trie.

3.5 Multibit Trie

Nodes of a trie can be compressed with a large number of bits, as the Patricia trie does, but rather using a constant number of bits, called *stride*. The multibit trie approach avoids the complexity of identifying the skip size and segment for each node. This approach also reduces the height of the trie by assigning a stride larger than one bit, and therefore, the search for a matching prefix would use a number of bits, equal to the stride size, at a time. In turn, the number of memory accesses is reduced. This approach is called a multibit trie [159]. In such a trie, prefixes are expanded to the levels that are proportional to the stride size, as Figure 3.5 shows it as a *n*-ary trie.

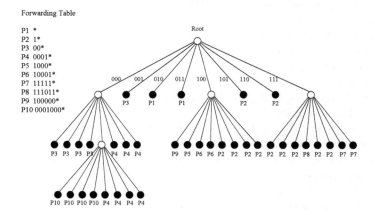

FIGURE 3.5
Multibit trie with a three-bit and constant stride.

The multibit trie may also be represented as a set of memory blocks, each holding all the children of a node. Figures 3.5 and 3.6 show an example of a

3-bit stride multibit trie. In these figures, the prefixes are pushed to the leaves when they fall into intermediate nodes. This approach is called multibit trie with leaf pushing. As the trie has a height of 7, the trie has three 3-bit levels, and prefixes have to be expanded to levels that are proportional to three: 3, 6, and 9.

Speed and memory amount. The multibit trie is faster than a binary tree for performing IP lookup. For strides of up to k bits, the trie's height is reduced from W, which is 32 bits for IPv4 (and 128 for IPv6), to $\lceil \frac{W}{k} \rceil$. Therefore, the search complexity is $O(W/k)$, where k can also be considered as the speedup factor. In this approach, the amount of memory used to store all prefixes of a forwarding table is larger than that used in a binary tree, as per the used prefix expansion, where in the worst case, a node is represented by 2^k children.

3.6 Level-Compressed Trie

The level-compressed trie [124], also known as LC trie, reduces a tree height and, therefore, the number of memory accesses needed to perform IP lookup. In addition, it reduces the required amount of memory to store the prefixes, in comparison with a multibit trie. The LC trie uses a combination of properties from the path-compressed, multibit trie, and disjoint tries. In fact, tree completion (as in a disjoint trie), bit expansion (as in a multibit trie), and path compression (as in a Patricia tree) techniques may be used to build an LC trie. This trie is a variable-stride multibit tree. It reduces the amount needed to build a tree by reducing unnecessary prefix expansions, by using variable stride sizes and path compression.

Figures 3.7, 3.8, and 3.9 show three steps in which an LC trie is built. First, a tree is completed, as Figure 3.7 shows.

Then, a stride size is selected to include a large number of prefixes at the selected prefix level (Figure 3.8). Once the stride is selected, the node is represented as a multibit trie.

After that step, the next compressed path is selected, as a new stride or as path compression. The main objective may be to reduce the tree height or memory. These steps are performed in the remaining uncompressed parts of the tree. Subtrees are compressed as a multibit or as a path-compressed trie separately from the compression of other subtrees. Figure 3.9 shows the resulting LC trie, where the stride size for the subtrees is three. Note that a stride size of four for the subtree at the (bottom) left would have reduced the number of memory accesses by one. However, at the expense using of more memory.

The largest number of memory accesses needed for this example tree is

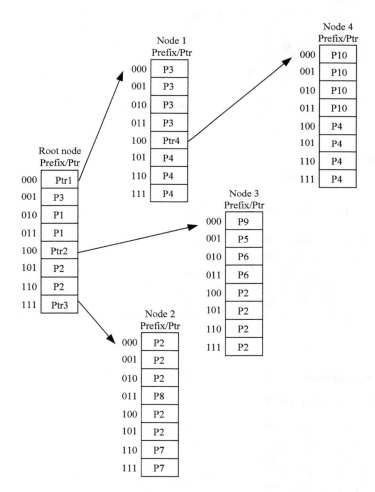

FIGURE 3.6

Multibit trie representation. Each node has 2^{stride} children.

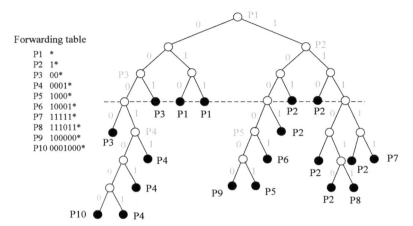

FIGURE 3.7
Phase 1 of LC trie construction: selecting level 3 and performing leaf pushing.

four as the root of the tree may need to be accessed for retrieving the stride size.

Speed and memory amount. The LC trie is very effective at reducing both memory size and number of memory accesses. As in the multibit trie, the stride is the number of bits that can be compared at once. The speed in which IP lookup is achieved depends on the stride sizes used. The lower the level where the prefixes are placed, the longer the stride can be achieved. The lookup complexity of the LC trie is similar to that of the multibit trie.

3.7   Lulea Algorithm

The Lulea algorithm, named after the university where it was developed, is a data structure that minimizes the amount of memory used to store NHIs of a forwarding table [49]. Lulea follows the representation of prefixes of a multibit trie but with fewer prefix levels and with a reduced number of replicated prefixes produced by expansion (i.e., leaf pushing). Lulea uses a codeword array, base index array, maptable, and NHI table. Each of these parts can be considered a memory block. The three first parts are used to determine which entry of the NHI table holds the matching prefix.

Lulea uses several strategies to reduce the memory amount: 1) a small number of prefix lengths (e.g., three in the original algorithm), 2) a prefix count, and 3) representing the leaves of a tree produced by expansion of prefixes, as disjoint subtrees, in a small number of combinations. Each of these strategies is described next.

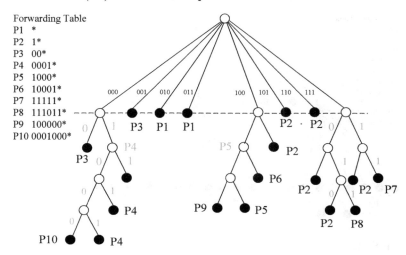

Forwarding Table
P1 *
P2 1*
P3 00*
P4 0001*
P5 1000*
P6 10001*
P7 11111*
P8 111011*
P9 100000*
P10 0001000*

FIGURE 3.8

Phase 2 of LC trie construction: multibit trie with stride 3.

3.7.1 Number of Prefix Levels

The Lulea algorithm is one of the first schemes that aims to represent all prefixes on a routing table on a few prefix levels. The selected levels are 16, 24, and 32. Any prefix with a prefix length smaller than 16 is expanded to that level, prefixes with prefix lengths between 17 and 23 are expanded to level 24, and similarly, prefixes with prefix lengths between 25 and 31 are expanded to level 32. The main level in the scheme is level 16. Levels 24 and 32 are accessed after level 16 is accessed. Figure 3.10 shows a general representation of the relocation of prefixes in the Lulea algorithm, for IPv4 tables. The description focuses on describing level 16 as this is used to build the maptable, which is reused for matching at levels 24 and 32.

3.7.2 Representation of Prefixes as a Bit Vector

To build the codeword and base-index arrays, and the contents of the maptable, prefixes in level 16 are represented as a bit vector. The bit vector is a string of bits, whose position represents the prefix value. The objective of the bit vector is to show the number of prefixes needed to be stored in the NHI. The bitmap has 2^{16} bits, numbered from bit 0 to bit $2^{16} - 1$. Here, each bit is dedicated to indicate whether its position holds a new prefix or not. For example, prefix 1000 0000 0000 0000* in binary, or 128.0/16 in decimal, occupies position 2^{15} in the bitmap with 2^{16} bits. This bit vector then represents prefixes from 0.0/16, as the bit in position 1, to prefix 255.255/16, as the bit in position 2^{15}.

A bit in the bit vector is the leaf of the binary tree built with the prefixes

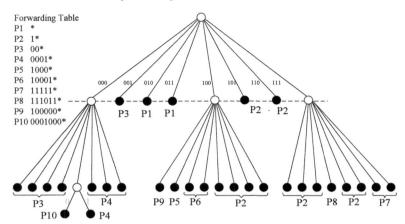

FIGURE 3.9
Final LC trie representation of the binary tree of Figure 3.2.

with lengths up to level 16. Figure 3.11 shows an example of how to set the bit values in the bit vector. This subtree has four levels, so it has 16 leaves; bits 0 to 15. Here, bit 0 is set to 1 as it is the first bit of a prefix expanded to four leaves. Bits 1 to 3 are also expansions of the same prefix, but since they are a duplicate of the prefix in bit 0, they are set to 0. Similar case is presented in bits 4 and 5. Because these two bits are expansions of the same prefix, bit 4 is set to 1 and bit 5 to 0. Bits 6, 12, and 13 are not expansions of a prefix but they are ancestors of prefixes, which have a longer prefix length. These bits are also set to 1. Therefore, nodes with no prefix and no descendants, as nodes that hold a duplicated prefix, are set to 0.

The number of bits set to 1 in the bit vector equal to the number of prefixes in the NHI table. By setting bits of duplicated prefixes to 0, the number of prefixes in the NHI table is kept small. For example, considering prefix length of four, if bits nodes 0001, 0010, and 0011 belong to the same prefix, the nodes are represented by the bits 1, 0, and 0, respectively (instead of representing them as 1, 1, and 1, respectively, as in a multibit trie). In this way, once a prefix, a bit of the bitmap, is accessed for lookup, the position of that prefix in the NHI table is addressed by the number of 1's in the bitmap of the bits with a position smaller than or equal to the matching prefix.

Figure 3.11 shows an example of the representation of nodes in a bitmap.

It should be clear now that what Lulea does is to count the number of ones from bit 0 to the bit x, where x corresponds to the matching prefix at level 16 in this case. However, storing the bit vector as described may not be efficient for counting all 1s, as multiple memory accesses would be needed. To overcome this issue, the scheme uses a codeword, base index array, and a maptable to store information about the bit-vector, and in turn, for the lookup process.

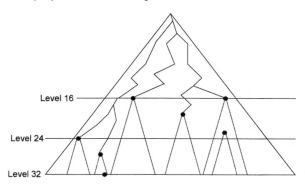

FIGURE 3.10
Prefixes are found at different levels and they can be projected onto the larger selected prefix level.

The bit vector is segmented into groups of small and equal number of bits, or *bit-masks*. The original Lulea algorithm uses 16-bit bit-masks. Therefore, level 16 comprises $\frac{2^{16}}{2^4} = 4096$ bit-masks. A codeword uses the same number of bits as a chunk and comprises two fields: 1) an *offset* and 2) a combination *code*. Therefore, the number of codewords in the codeword array is 4096, where each codeword corresponds to one and only one bit-mask. The offset field of codeword i indicates the number of 1s in bit-masks $i - 3$, $i - 2$ and $i - 1$, where i is the bit-mask of the matching (prefix) bit. The offset records the accumulated the number of ones in each set of four bit-masks. The offset of bit-mask 0 is 0, as the offset is 6-bit wide. However, the count of 1s must include all those from bit-mask 0 to bit-mask 2 (three bit-masks) in the set of four. The counting of 1s in the last bit-mask is stored in the maptable.

To add the 1s up to bit-mask $2^{12} - 2$, a *base index* is used. Figure 3.13 shows an example of the use of the code, offset, and bit index. The length, in number of bits, of the base index is calculated by considering the largest number of 1s that may accumulate for the last possible bit-mask in the bit vector, or $2^{16} - 16$ 1s for level 16. Therefore, the base index is 16 bits long (i.e., $\lceil \log_2(2^{16} - 16) \rceil$. Because offset counts the number of ones in three bit-masks, a base index is allocated for each set of four consecutive bit-masks (this reduces the number of base indexes). Therefore, the base index array has $\frac{2^{16}}{2^6} = 1024$ base indexes. Therefore, the MSB 12 bits of a packet destination address are used to access a codeword and 10 bits for a base index.

Figure 3.13 shows the codeword and base index arrays.

As an example, let's consider that a packet address matches the bit position pointer by the arrow in Figure 3.12. The arrow points to the bit corresponding to prefix 00000000 01011101* (or bit 92), which is in Bit-mask 5 (the sixth bit-mask from the left). Therefore, base index 1 counts the number of 1s in Bit-masks 0 to 3, offset 5 counts the number of 1s in Bit-mask 4, and code

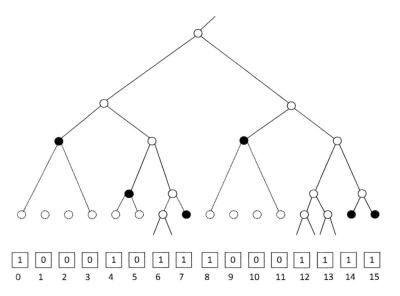

FIGURE 3.11
Prefixes at levels lower than the target level can be projected onto the target
level.

in Bit-mask 5 counts the number of 1s from the beginning of the bit-mask
to the bit corresponding to the prefix. The total number of ones is 22, which
corresponds to prefix **v**.

3.7.2.1 Code Field and Maptable

The code field indicates the combination and position of the 1s in a bit-mask.
The code is used to address an entry in the maptable. This entry provides this
information by showing the accumulated number of 1s for each bit position in
the corresponding bit-mask. Figure 3.14 shows the maptable for our example.
The table has as many rows as the number of different codes and as many

FIGURE 3.12
Example of the 1 bits counted by the different fields.

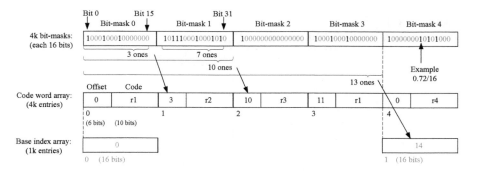

FIGURE 3.13
Example of the values of offset and base index fields.

columns as the number of bits in a bit-mask. Therefore, the code and the four LSB bits in a packet's destination address are used to address the rows and columns of the table, respectively. In general, the $\log_2$(bit-mask size) LSB bits (e.g., 4 bits for level 16) of the string of packet address bits used for lookup (e.g., 16 bits for level 16) are used to address the maptable column.

(Code) address	0	1	2	3	4	5	6	7	8	9	10	11	12	13	14	15
575																
								•								
								•								
								•								
(r4) 3	1	1	1	1	1	1	1	1	2	2	3	3	4	4	4	4
(r3) 2	1	1	1	1	1	1	1	1	1	1	1	1	1	1	1	1
(r2) 1	1	1	2	3	4	4	4	4	5	5	5	5	6	6	7	7
(r1) 0	1	1	1	1	2	2	2	2	3	3	3	3	3	3	3	3

FIGURE 3.14
Example of maptable for bitmap of Figure 3.12.

Table 3.4 shows the NHI table used in Lulea to identify the matching prefix. The number of prefixes (and entries) in this table is equal to the number of 1s in the bitmap.

3.7.2.2 Code Field

The authors of the Lulea algorithm found that the number of combinations in a 16-bit bit-mask is smaller than 2^{16}. As the expansion of prefixes is similar to completing a tree, where each node has either two or no children (see the disjoint trie), the number of actual combinations that can be produced in prefix expansion is smaller than 2^l, where l is the prefix level. Specifically, the number of combinations are

TABLE 3.4
Next hop index table of bit vector in Figure 3.12.

Memory address	Prefix Id (NHI)
1	a
2	b
3	c
4	d
5	e
6	f
7	g
8	h
9	i
10	j
11	k
12	l
13	m
14	n
15	o
16	p
17	q
18	r
19	s
20	t
21	u
22	v

$$a(0) = 1,$$

$$a(n) = 1 + a(n-1)^2 \tag{3.1}$$

for bit-masks with 2^n bits. For example, if the bit-mask is 16-bit wide, $n = 4$ as $2^4 = 16$, and $a(4) = 677$ (an improvement to reduce this number of codes has been considered [49]). After considering the all-0s combination $(677 + 1)$, to represent 678 combinations, we need only 10 bits ($\lceil \log_2(678) \rceil = 10$) instead of the original 16 bits. Then, each combination of 0s and 1s in a bit-mask with 16 bits is represented as one of the 2^{10} combinations. This shows that the code provides information of the combination of 1s and 0s in a bit-mask but with a smaller number of bit than those in the bit-mask. In general, the size of the code, $|code| = \lceil \log_2(a(n)) \rceil$, in number of bits.

Figure 3.15 shows the combinations of complete tries with $n = 1$ and 2. These examples shows that the possible combinations are only two and five, respectively.

Disjoint tree with up to 1 level (n=1)

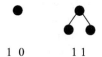

1 0   1 1

Disjoint tree with up to 2 levels (n=2)

1 0 0 0   1 0 1 0   1 1 1 0   1 0 1 1   1 1 1 1

FIGURE 3.15
Combinations of complete trees for one and two bit Lulea codes.

3.7.2.3   Lulea Matching Process

The lookup process in Lulea uses the destination IP address to identify the corresponding NHI in the NHI table. For this, the MSB 12 bits of the address are used to find the corresponding bit-mask and the MSB 10 bits for accessing the corresponding base index. The LSB 4 bits of the level (16 in this case) are used as partial address to the maptable. Figure 3.16 shows the a summary of the bits needed from the packet destination address to access the code and NHI tables.

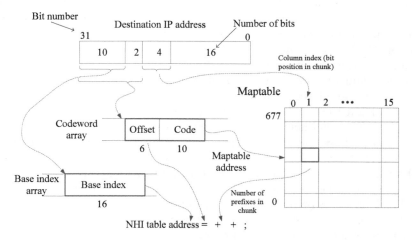

FIGURE 3.16
Lulea search process.

For example, let's consider IP lookup is performed for a packet with destination address 0.72.0.0 in the bit vector used as an example (Figure 3.13). Using the MSB 16 bits part of the address, 0.72, matches Codeword 5 and Base index 1. Codeword 5 has offset = 0 (as it is the first bit-mask of a set of four), the code is r4 (or row 4 of the maptable), and base index = 14. The maptable entry addressed by row 4 and column 9 stores the number 2, so the address in NHI=0+14+2=16. The reader can count the number of 1s to where the *Example* arrow in Figure 3.16 and this is the same number. The 16th NHI prefix is matched and it may provide the next hop address for this packet. This prefix corresponds to prefix **p** in Table 3.4 and Figure 3.12.

For level 16, Lulea performs IP lookup in three memory accesses (where the codeword and base index arrays are accessed simultaneously). A similar process is performed for other larger prefix levels. A corresponding number of memory accesses is added to the lookup time if the matches continues to the other levels. Forwarding table updates are costly in this scheme, as codewords, base indexes, maptable, and NHI table may need to be all updated.

3.8   DIR-24-8 Scheme

The DIR-24-8 scheme aims to represent prefixes in two levels (or lengths) [75], similar to the Lulea algorithm. This scheme performs controlled prefix expansion to represent all prefixes on either level 24 or 32. The adoption of two levels greatly simplifies the implementation of the IP lookup engine. The scheme uses a memory block for each level. Level 24 uses a memory large enough to represent all possible prefixes. All prefixes with length smaller than 24 are expanded to that level. Prefixes between levels 25 and 31 are expanded to level 32. As the number of prefixes at level 32 is too vast, only a few 256-address intervals are represented. Each of these intervals, or subtries, are the descendant of a prefix at levels 25 to 30. Figure 3.17 shows the idea of representing these intervals at level 32. The total number of intervals at level 32 depends on the number of prefixes larger than 24 bits in the forwarding table. Because it is difficult to estimate the memory needed for an unknown number of prefixes, the level-32 memory is given a fixed size, and therefore, the number of prefixes is limited to those that fit in that memory size.

DIR-24-8 uses two memory blocks, called TBL24 and TBLong, where the first one stores prefixes on level 24, or pointers directing the search to level 32, and TBLlong stores prefixes on level 32. Figure 3.18 shows these two memories and the supporting logic. An entry (memory location) holds a prefix bit to indicate whether the entry is a prefix. If the bit is set to one, the stored information is the NHI; otherwise it is a pointer to TBLlong.

Speed and memory amount. DIR-24-8 performs IP lookup in up to two memory accesses (if the prefix is not matched at TBL24, it is matched in TB-Long). The lookup complexity of DIR-24-8 is O(1) as the search on TBL24 is

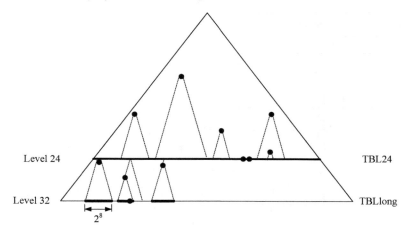

FIGURE 3.17
Prefix levels in DIR24-8 scheme and the prefix intervals created by prefix expansion.

a single memory access, where the 24 MSB bits of a packet destination address is used as memory address, and the process is similar in TBLong, however, using instead 32 bits. The number of maximum number of pointers is 2^{12} and it may need to copy several replicates of NHI for expanded prefixes. The table update is simple as the small number of levels simplifies the identification of entries to be updated but it may require accessing several entries (as a by product of prefix expansion). As can be deduced from Figure 3.18, the total memory amount for this scheme is $(2^{24} + 2^{20})W$, where W is the number of bits used in an entry. The entry stores the prefix bit and the NHI. Figure 3.19 shows an example for interval 123, to where some prefixes are expanded.

3.9   Bloom Filter-Based Algorithm

Bloom Filter Theory. A Bloom filter is a space-efficient data structure for membership queries [19] and is widely used in various applications because of its compactness and simplicity. A Bloom filter is an m-bit array that contains abstract information about a set. For a given set $S = \{x_1, x_2, \cdots, x_n\}$, programming a Bloom filter starts from initializing every bit of the filter to zero. For element x_j, where $1 \leq j \leq n$, k hash indices $h_i(x_j)$ for $1 \leq i \leq k$ are obtained, where $0 \leq h_i(x_j) < m$. To program element x_j into a Bloom filter, the k bits indicated by $h_i(x_j)$ are set to one. This procedure is repeated for every element included in the set S.

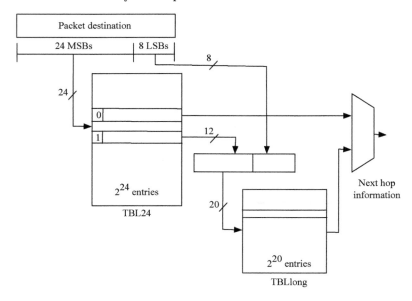

FIGURE 3.18
Implementation of DIR-24-8

Querying for the membership of a given input y also uses the same k hash functions, $h_i(y)$ for $1 \leq i \leq k$. The k bits indicated by the hash indices are checked. If all k bits are one, y is considered a member of S. If any of the k bits are zero, y is definitely not a member of S.

Bloom filters may have false positives, but no false negatives. For n elements, the false-positive rate f of an m-bit Bloom filter is calculated as Equation 3.2 [118].

$$f = \left[1 - \left(1 - \frac{1}{m} \right)^{kn} \right]^k \approx \left(1 - e^{-kn/m} \right)^k \qquad (3.2)$$

The optimal number of hash functions for an m-bit Bloom filter of n elements can be calculated as follows [118]:

$$k_{opt} = \frac{m}{n} \ln 2 \qquad (3.3)$$

Figure 3.20 shows an example of a Bloom filter. The Bloom filter is programmed by $S = \{x_1, x_2, x_3, x_4\}$, and queried by inputs y and z. The Bloom filter produces a negative result for y and a false-positive result for z.

3.9.0.4   IP Address Lookup Using a Bloom Filter

In a binary trie, a node cannot exist without ancestor nodes, except for the root. This principle is used by a Bloom filter to improve the performance in a

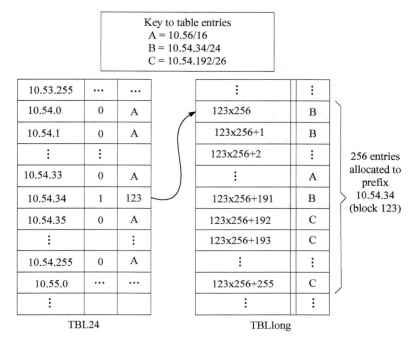

FIGURE 3.19
Example of interval 123 with expanded prefixes on TBLong.

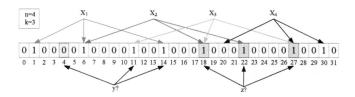

FIGURE 3.20
An example of a Bloom filter.

search on a binary trie [118]. The Bloom filter may be implemented using on-chip memory, while a trie is usually implemented in off-chip memory because of its larger memory requirements. The Bloom filter is used to reduce the number of off-chip memory accesses by determining the best matching level holding the longest matching prefix for a given packet destination address.

In this approach, nodes of a binary trie are programmed in an on-chip Bloom filter and stored in an off-chip hash table. The Bloom filter is queried, while linearly increasing the access level, until a negative result is produced. The negative result means that there is no node detected in the current and longer levels. Therefore, the level of the last positive outcome is the candidate level storing the best matching prefix. Using the prefix with a length equal to the candidate level, the off-chip hash table is probed to verify the result. Because the search procedure of this approach starts from the candidate level, the matching process may finish right after the matching prefix is found. If a matching node is not found because of the Bloom filter's false positive, the search procedure goes to a smaller level. This process is called *back-tracking*.

The number of back-tracking processes depends on the rate of false positives. Assuming that the best matching prefix is on level L, if a false positive at level $L+1$ and a negative at level $L+2$ are produced, a single back-tracking process takes place. Similarly, if false positives on levels $L+1$ and $L+2$ and a negative outcome occur at level $L+3$, two back-tracking processes take place. Hence, the probability of the number of back-tracking processes are calculated by Equation 3.2, where b_j represents that back-tracking has occurred j times and f_i is the false-positive rate of a node in level i.

$$Prob(b_j) = \prod_{i=L+1}^{L+j} f_i \tag{3.4}$$

The probability that multiple back-tracking processes occur rapidly decreases because this probability is the product of the false-positive probability for each level. However, even though a positive result occurs, the positive result does not guarantee that the best matching prefix is on the level, because empty internal nodes may exist. If an internal node is encountered, the back-tracking process also takes place. Precomputing the best matching prefix, and storing the corresponding information for each internal node can easily solve this problem. Figure 3.21 shows the IP address lookup algorithm using a Bloom filter for the example forwarding table in Table 3.5. It is assumed that each gray-colored node in Figure 3.21 inherits the longest matching prefix from its direct ancestor prefix node.

TABLE 3.5
Example of a forwarding table.

IP prefix	prefix ID
00*	P0
010*	P1
1*	P2
110101*	P3
1101*	P4
111*	P5
11111*	P6

Note that a hash entry is accessed either for the level with the last positive result when the Bloom filter produces a negative or by a back-tracking process

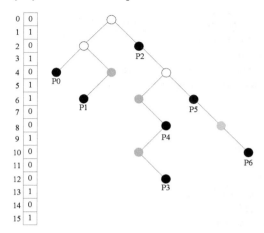

0	0
1	1
2	0
3	1
4	0
5	1
6	1
7	0
8	0
9	1
10	0
11	0
12	0
13	1
14	0
15	1

FIGURE 3.21
Improving trie search using a Bloom filter.

when the Bloom filter produces a false positive. If a node at level L has both children, the result is neither a negative nor a false positive at level $L + 1$. Hence internal nodes with both children (denoted with dotted lines in Figure 3.21) are not necessarily stored in the hash table, while every node in a trie is programmed in the Bloom filter.

Algorithm 1 describes the operation of this lookup scheme. For example, the search procedure for the input address of 110100 is as follows: The Bloom filter is queried for lengths 1 through 5 and it produces positive results. At level 6, the Bloom filter produces a negative result. Hence the hash table is accessed at level 5 using the prefix of 11010, which is the level of the last positive result, and the search is over by returning the precomputed longest matching prefix $P4$.

3.10 Helix: A Parallel-Search Lookup Scheme

As discussed in the previous sections, IP lookup schemes based on binary trees require multiple memory accesses and that makes it challenging to use them in implementations for routers with high-speed ports. To overcome this, Helix is a scheme that performs both IP lookup and table updates for IPv4 and IPv6 prefixes in a single memory access [145]. Helix achieves this lookup speed using very small amounts of memory. Helix uses a parallel prefix search [49, 140, 191] at the different prefix lengths and the helicoidal properties of binary trees. These properties are based on observations that if a binary tree

---

Algorithm 1 Address Lookup Procedure in Bloom Filter-based Approach

---

Function: Search dstAddr
find the BML using Bloom filter
for $i \leftarrow shortestLen\ to\ longestLen$ **do**
  if $queryBF(dstAddr,\ i)\ is\ positive$ **then**
    $BML \leftarrow i$;
  else
    break;
  end if
end for
access hash table
for $i \leftarrow BML\ to\ shortestLen$ **do**
  $node \leftarrow queryHT(dstAddr,\ i)$;
  if $node\ exists$ **then**
    return node.info;
  else
    continue; // back tracking
  end if
end for

---

is considered with plasticity features, it could be "folded" to make a smaller tree. The helicoidal properties of a binary tree allow folding the tree (i.e., information is not lost after the tree is folded).

Parallel search means here that Helix searches for a matching prefix in all different prefix tables, simultaneously. Parallel search has been of interest in several schemes [49, 140]. Among the matched prefixes at the different levels, the longest prefix is selected.

Figure 3.22 shows the Helix's lookup process for IPv4 prefixes, as an example. In this figure, there is a prefix table for each nonempty prefix level. A matched prefix, if any, is output by the corresponding level table(s), and the selector selects the longest matching prefix.

A nonempty prefix table is one that has one or more prefixes. Prefixes are kept in their original levels; prefix lengths remain unchanged. In this scheme, there is a prefix table for each nonempty prefix level, where each table is implemented as a memory block. Note that keeping the prefixes in their original length has the following benefits: 1) the number of prefixes in the binary tree does not increase and 2) it is simple to make table updates.

Small tree levels may be stored without any change (e.g., level 8 may be a memory with 256 entries and the level-8 prefixes are stored as they appear in a routing table). Large levels may need to be processed further.

A forwarding table has as many different NHIs as the number of the ports, k, of the router hosting the table. Table 3.2 shows an example of a forwarding table, which is used as an example throughout this section to describe the proposed scheme and the helicoidal properties. Herein we use the prefix

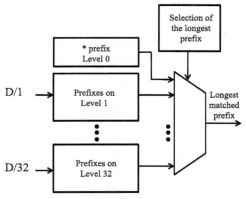

D: Destination IP address of incoming packet

FIGURE 3.22
Parallel-level lookup process.

identification label instead of the NHI for simplicity. Figure 3.23 shows the prefixes listed in Table 3.10 as a binary tree. The height of this tree is six, and the tree is also referred to as having six levels. Therefore, prefix length y of prefix x is referred to as the level of the tree where the prefix is located. In this example, prefix **a** ($x = 0^*$) is located at level 1 ($y = 1$), and prefix l ($x = 111111^*$) is located at level 6 ($y = 6$).

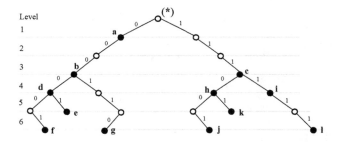

FIGURE 3.23
Binary tree of the prefixes in Table 3.10.

This binary tree, as shown in Figure 3.23, has six levels but only five levels hold prefixes; levels 1, 3, 4, 5, and 6. Therefore, five prefix tables are provisioned. The table for level 6 requires 64 memory locations and holds prefixes **f**, **g**, **j**, and l.

The memory length increases exponentially as the level increases [145]. Therefore, the memory length is the dominant factor in the determination of the amount of used memory [11, 124].

Helix minimizes the number of bits used to address the location of prefix

TABLE 3.6
Prefixes of an example forwarding table.

IP Prefix	Prefix ID
0*	a
000*	b
111*	c
0000*	d
00001*	e
000001*	f
000110*	g
1110*	h
1111*	i
111001*	j
11101*	k
111111*	l

x in a prefix table by splitting the prefix into two parts: the MSB portion or *family root (FRx)* of x and the LSB portion or *descendant bits (Dx)* of x. Figure 3.24 shows these two portions of bits in a prefix, separated between bits of level T and $T + 1$. FRx is stored with the corresponding NHI in the location addressed by Dx.

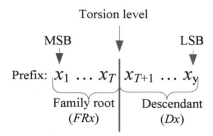

FIGURE 3.24
Family root and descendant bits in a prefix after a torsion.

The line between FRx and Dx in the figure, called the torsion level (L_T), indicates the bit where the original prefix is split. The selection of L_T determines the size of FRx and Dx, in number of bits, and subsequently the size of the memory amount used for prefix level v.

As an example, let's consider level 6 of the example binary tree after applying a torsion on level 3 ($L_T = 3$). In this case prefixes on level 6, **f**, **g**, **j**, and **l**, are represented as $FRf = 000$, $FRg = 000$, $FRj = 111$, and $FRl = 111$, while $Df = 001$, $Dg = 110$, $Dj = 001$, and $Dl = 111$. Since Dx is used to address the location of prefixes, we see that multiple prefixes may share the memory location. In our example, prefixes **f** and **j** share the same location. These prefixes are said to be *conjoined*. Figures 3.25(a) and 3.25(b) show a prefix table with **f** and **g** sharing the same memory location.

(a)

Prefix	FRx	Descendant
f	000	001
g	000	110
j	111	001
l	111	111

Same descendant bits

(b)

Dx	Entry 1		Entry 2	
	FRx_1	x_1	FRx_2	x_2
000				
001	000	f	111	j
⋮	⋮	⋮	⋮	⋮
110	000	g		
111	111	l		

FIGURE 3.25
(a) FRx and Dx parts of prefixes at level 6 after $L_T = 3$. (b) Reduced prefix table of level 6.

Figure 3.26(a) shows the same tree and additional nodes to help visualize the representation of a torsion. The additional nodes are shown with a dashed-line circumference. Figure 3.26(b) shows the tree after the torsion.

This figure also shows the torsion on level 3 of the example binary tree, the resulting subtrees below the torsion level, and the representation of conjoined prefixes on level 6 of the resulting tree. The nodes below L_T form subtrees rooted to their string of ancestor nodes; from the root (*) on level 0 to L_T. This string of bits is FRx. The remaining string of bits is Dx.

For every prefix level, a torsion level is selected as to reduce the tree height such that the number of entries would fit in a memory block and would be fetched in a single memory access. The use of the Helix data structure then reduces the size of a binary tree, and in turn, it achieves optimum lookup speed and small memory use. Helix makes binary trees perform comparably to lookup engines based on expensive ternary content addressable memories (TCAMs).

3.11 Ternary Content-Addressable Memory

The implementation of the schemes presented above is based on RAM. However, associative memory is a memory that instead of matching the address input it matches the content. The use of this memory may bring additional benefits, such as embedded multibit search for matching prefixes and higher resolution speed. Such memory is called content addressable memory (CAM).

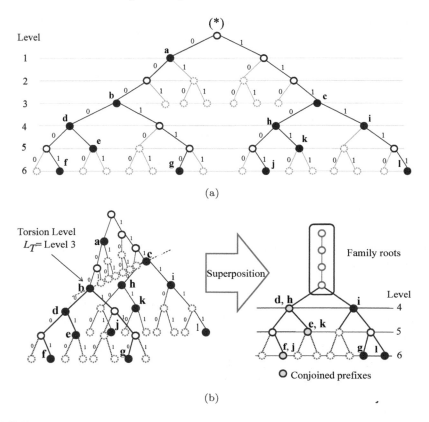

FIGURE 3.26
(a) Example of the binary tree in Figure 3.23. (b) Levels 4, 5, and 6 form two subtrees after a torsion, which are superposed.

A version of a CAM where masks can be used to mark bits as a "don't care" value, or a third value, is called Ternary CAM (TCAM).

In short, TCAMs receive an input, the content that needs to be matched, and it outputs secondary information associated with it. The associated information could be the NHI of a prefix, or the address of where the NHI is in an additional RAM block. Prefixes of different lengths may be stored in a TCAM. Therefore, when a packet destination is input as the content to a TCAM, multiple prefixes with different lengths may match, but only the longest prefix is of interest.

To output a single match, a TCAM matches the first content input into it. That is, longer prefixes are stored in the highest priority positions (or input first), and they are followed by prefixes with immediately shorter lengths, and so on. This storing process for a TCAM solves the matching priority operation required by IP lookup, but also makes it complex to update changes in prefixes or routing tables [153]. Figure 3.27 shows a basic example of a TCAM storing prefixes. In the TCAM, there is a priority encoder that assigns selection priority to the prefix on the top. Therefore, longer prefixes are placed at the top of the memory. Once the prefix is matched, the information associated with the entry or the ID of the prefix may be output. In the example the figure shows, packet destination address 133.210.111.7 matches prefix 133.210.111/24, which is associated with next hop address 113.45.22.14. This address is then output by the TCAM.

IP lookup by a TCAM is achieved in a single memory access, which is the optimum lookup time. In the lookup process, when a packet arrives, the destination address is input to the TCAM and all the contents of the TCAM are compared to it.

With these features, TCAMs are a practical alternative to building IP lookup engines [154, 158, 189].

Although the lookup speed of TCAMs is very fast, there are still some issues that need to be considered. One of them is that all entries are compared at once. This comparison of a large number of prefixes requires a large number of comparators. The additional logic takes some room and limits the number of entries that can be stored. Also, access to all memory elements and comparison of their content consume power. Therefore TCAMs are then power-demanding systems.

Several schemes have been developed to overcome some of these issues [133, 189]. However, much of the complexity on addressing these issues is that TCAMs are integrated systems.

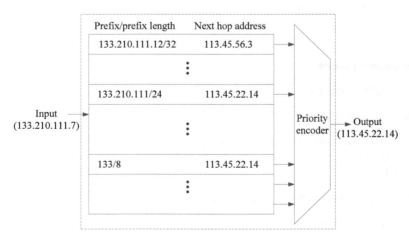

FIGURE 3.27
Example of prefixes in a TCAM.

TABLE 3.7
Forwarding table for Exercise 1.

IP Prefix	Prefix ID
*	P1
1*	P2
01*	P3
000*	P4
0110*	P5
0111*	P6
1001*	P7
1101*	P8
01001*	P9
11111*	P10
111000*	P11

3.12   Additional Readings

Search for simplification of IP lookup complexity has captured the interest of the research community for many years. The number of literature works is quite extensive. Therefore, there is a very long list of surveys of IP lookup algorithms for the interested reader [28, 152, 162, 179]. Other books may also list several schemes [29, 175].

3.13   Exercises

1. Build the binary tree representation from prefixes in Table 3.7.

2. From the forwarding table in Figure 3.7, show it as a disjoint tree, and indicate the largest number of memory accesses that may be needed to match a prefix.

3. Show a path-compressed tree (Patricia tree) of Table 3.7.

4. Indicate the largest number of memory accesses to match a prefix in Exercise 3.

5. Show a Lulea bitmap for level 4 of the forwarding table (or binary tree) of Table 3.7. How many bits can you use for index and code word codes?

6. Show a level-compressed tree (LC-Trie) of prefixes in Table 3.7 with initial stride of 3.

7. Show the multibit trie of the prefixes in Table 3.7 with stride 3. What is the largest number of memory accesses needed for this table?

8. Show the binary tree of the following forwarding table: P1 (*), P2 (1*), P3 (00*), P4 (01*), P5 (1101*), P6 (00101*), P7 (00111*), P8 (111010*), P9 (1110111*), P10 (11101010*).

9. Show the disjoint tree of the forwarding table in Exercise 8.

10. Show the path-compressed tree of the forwarding table in Exercise 8.

11. Show the Lulea data structure, including codeword and base index arrays, maptable, and NHI table for level 7 of the forwarding table in Exercise 8. Use 8 bits as bit-mask size.

12. Find the torsion level, other than level 7, that would require the smallest amount of memory to represent the prefixes using the Helix data structure of the list prefixes below. Indicate the memory size (in length and width) and the total amount (in bits) if 8 bits are used to represent the NHI of a prefix. List of prefixes: P1 (0000001), P2 (0000111), P3 (0011011), P4 (0100111), P5 (0110011), P6 (0111101), P7 (1000111), P8 (1010001), P9 (1011011), P10 (1100111), P11 (1111011), and P12 (1111101).

4

Packet Classification

CONTENTS

Packet classification is a function that routers or other network appliances perform to identify a packet for asserting the type of service that the router will perform on the packet. This function must be performed accurately and fast. This chapter discusses packet classification and different schemes to perform it fast and efficiently.

4.1 Introduction to Packet Classification

Once entering an edge router, packets may be classified according to the required service or protective action. A service may consist of allocating a portion of link bandwidth or policing the burstiness of a stream of packets of a flow. Here, we call a flow to the set of packets with the same destination-source IP addresses. This flow definition may be made more specific if it includes other information from the transport- and network-layer packet headers, such as port numbers, transport protocol number, or flags.

A common possible action on a flow is fire walling, where some flows are allowed to enter a network and others, considered as threats, are dropped. Figure 4.1 shows an example of a network using packet classification and applying different actions on different flows. This figure shows a network, identified as 128.235/16, and other neighbor networks connected to it. The 128.235/16 network has four routers, two of which are edge routers (R1 and R4). Edge routers provide internetwork connectivity. Router R3 is a router interconnecting R1 and R4, and R2 provides access to two service servers: a File Transfer Protocol (FTP) and World Wide Web (WWW) servers. The router uses a firewall for network protection. The traffic coming in, out, and passing through the network may be regulated by the edge routers.

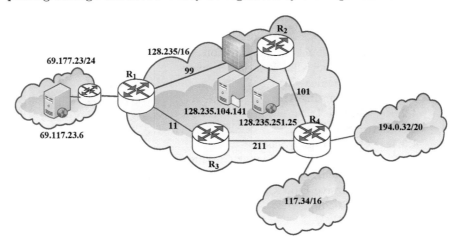

FIGURE 4.1
Example of a network where packet classification is used.

To indicate a router how to manage incoming traffic, a system administrator defines rules for each router. A rule is a set of conditions associated with an action. The conditions are information carried by the packets in their headers and the actions are taken if the header information of a packet matches the set of conditions. This is, if a packet of a flow matches a rule, the action is performed on each one of the packets of this flow.

The set of rules in a router (or other network device) is called a classifier. Rules are defined with different priorities by the system administrator of the network (or autonomous system); for example, rules are sorted in the classifier according to their priority. The top priority rule is placed as Rule 1, the second priority rule is placed as Rule 2, etc. The definition of priorities is then indicated by the order in which rules are listed. Table 4.1 shows an example of a classifier with the rules used in the network of Figure 4.1. The classifier has five rules. Rule 1 is used to discard any packet with destination to 128.235.104/24 for destination port (transport layer) 80 (or HyperText Transport Protocol, http). Rule 2 is used to forward packets from 194.0.32/20 that uses the Transport Control Protocol (TCP) to Router 3 (R3). Rule 3 is used to forward packets destined to 128.235.251/24, using TCP as transport protocol to communicate with a web server (port 80), to Router 2 (R_2). Rule 4 is used to allow User Defined Protocol (UDP) traffic to enter destination 69.114.23/24, with source ports in the range from 17 to 99 as long as traffic comes at a rate equal to or smaller than 5 Mbps. Rule 5 permits TCP traffic to 194.0.32/24, and destined for a range of (transport layer) ports from 21 to 88, to pass through.

TABLE 4.1

Example of classifier for Figure 4.1.

Rule	L3 Src Add	L3 Dst Add	Protocol	L4 Src Port No.	L4 Dst Port No.	Action
(1)	*	128.235.104/24	*	*	80	Discard
(2)	194.0.32/20	*	TCP	*	*	Forward to R3
(3)	*	128.235. 251/24	TCP	*	80	Forward to R2
(4)	*	69.114.23/24	UDP	17-99	*	Allow ≤ 5 Mbps
(5)	*	194.0.32/24	TCP	*	21-88	Permit

As discussed above, a match in packet classification must be an exact match in each of the considered fields. Yet, a rule may specify not exactly a particular value but a range of them for a given field. For example, the address field in the classifier may indicate a single network (IP) address or a range of them (a prefix and the corresponding prefix length). The example above (Table 4.1) shows range of prefixes in the source and destination address fields.

At first sight, it may seem that matching a field, particularly those related to IP addresses, may be the same as IP lookup. But, packet classification is in fact different. In packet classification, there are many rules in a classifier. These could be tens, hundreds, or thousands of rules. These rules may be different for every single autonomous system (AS), network, or router. Moreover, a packet may match not only one rule but several of them. To differentiate one rule from another, rules are then assigned a weight [175] or priority. Therefore, to explicitly list the priority of rules in a classifier, rules are presented with a number that indicates the priority. For example, Rule 1 is the one with the highest priority and Rule 2, the rule with the second highest priority. Therefore, the difference between IP lookup and packet classification is that packet classification performs exact matching to the highest priority rule while IP lookup performs the longest (prefix) matching.

This chapter presents different packet classification schemes using different principles. They are sorted in three categories: 1) binary tries, 2) geometric representation, and 3) heuristic schemes. A Binary trie are used to represent string of bits of different protocol header fields and is mostly used to represent prefixes (or ranges of network addresses). The geometric representation schemes use the product of the information of two fields as an area (or intersection) [160]. The heuristic scheme presented in this chapter is a variation of a crossproduct-based scheme. The schemes are mainly described in three different components: 1) how to build the data structure of the classification scheme, 2) how to perform matching, and 3) advantages, disadvantages, and matching complexity.

4.2   Trie-Based Packet Classification Schemes

Packet classification schemes based on binary tries use similar principles to represent rules as it is done in binary tries used for IP lookup. In a classifier, a rule may define a specific address or a range of addresses, which is expressed as prefix with a prefix length (suffix), as in IP lookup. The representation of an address field may take the form of a binary tree. The tree starts from the root (or *), which is the complete range of addresses. This is often called the "wildcard" value, as it addresses all prefixes.

4.2.1   Hierarchical Trie

A method directly derived from binary trees for IP lookup is known as Hierarchical tree [163]. This is suitable for representing the two network-layer address fields, the source-address field (SF) and the destination-address field (DF), to find which rule is matched by the addresses in a packet. The construction of the tree uses these two sets of addresses. In the following example, the construction of the tree represented by an example two-field classifier starts with DF. The prefixes in DF are represented as a binary tree, starting from the MSB bit. Figure 4.2 shows the hierarchical tree of a two-field classifier.

After constructing the binary tree from DF, SF is used with each DF prefix as a branch continuation in the area marked as the SF zone in the figure. The branch continuation is linked by a pointer directed from DF to SF, as the figure shows in a straight dashed line. For example, prefix of rule R2 in DF, 01*, is linked to the SF prefix, 01*, through a pointer (dashed line). Since this is the only SF prefix linked to DF prefix 01*, this is the only part of the tree in the SF zone of the tree. This process is performed for all DF and SF prefixes.

Nodes that are prefixes in SF mark the match of a rule (indicated by the black nodes in the figure). Therefore, if a packet matches a rule, the destination of the packet matches DF bits and the source address of the packet matches SF bits. The matching of a complete prefix is performed by traveling from the MSB to the LSB; from the root (*) to a leaf. After reaching the branches of the DF and SF that match most bits from the addresses of the packet, the matching rules are noted. If Rule 1, which is the rule with the highest priority, is not matched in that search, then the matching process continues with the DF branch containing a shorter matching prefix and continued in the SF zone.

As an example of the matching process in a hierarchical tree, let us consider that an incoming packet has the (binary) destination address 000... and source

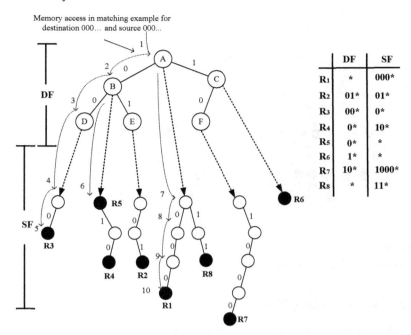

FIGURE 4.2
Example of a hierarchical tree representation.

address 000... The dotted arching arrows in Figure 4.2 show the search path followed according to the packet information. The destination address of the packet is used to find the matching path on the DF zone and the source address in the SF zone. As the figure shows, the first path on DF matches the first two bits of the destination address, 00, or node D, and then it continues in SF, matching bit 0 from the source address, matching R3. However, the branch 0 in DF, or node B, also matches the destination, and then the branch in SF linked to B is explored. Here, only the root (i.e., wildcard) of the SF branch matches, which is R5. As there is a rule associated with the wildcard in DF, or node A, the branch in SF linked to that node is also explored. In this linked branch, the source address matches all three bits of the branch, 000, and as a result R1. In summary, the packet matched rules R1, R3, and R5. Since R1 has the highest priority, R1 is the final result. As the figure shows through the dotted arrows, the matching process requires 10 memory accesses, where one memory access is a visit to a node.

In this approach, each visit to a node is a visit to a memory location (or memory access). Therefore, this scheme requires multiple memory accesses or in the order of $O(W)$, where W is the maximum length of a field (e.g., SF prefix plus DF prefix) in the classifier.

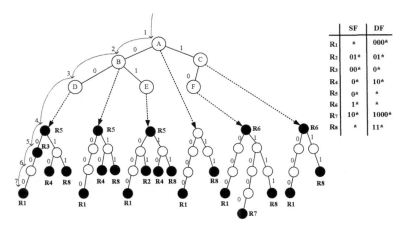

FIGURE 4.3
Set pruning tree of two fields.

4.2.2   Set Pruning Trie

The hierarchical tree is rather a simple method to store a classifier and find the matching rule. It takes several memory accesses to find the final match as all matching branches of the tree must be considered according to the information of the packet in process. As discussed before, the larger the number of memory accesses needed, the smaller the rate in which packets can be classified.

The Set Pruning trie is a method to reduce the number of memory accesses needed to perform packet classification. To achieve this smaller number of memory accesses, the subtrees in the DF zone rooted to shorter SF prefixes are replicated on subtrees (prefixes) in the DF zone of longer SF prefixes. For example, a DF subtree of the SF prefix 0* would be replicated to the subtrees of SF prefixes 00* and 01*. The DF subtrees of the SF prefix * would be replicated in all other branches, and so on. Figure 4.3 shows the resulting Set Pruning trie of the previous example.

The matching process in the Set Pruning trie is performed similarly as in the hierarchical tree, except that exploration of other matching subtrees becomes unnecessary, and therefore, it is not performed. This is because copies of the matching subtrees are now available in the longest branches. The result is a smaller number of memory accesses in the matching process. As in the previous example, a packet with destination 000... and source address 000... would now go through a single paths from the root to a leaf, as Figure 4.3 shows in dotted arched lines. In this example, the number of memory accesses is reduced from 10 to 7.

The storage and matching complexity are $O(N)$ and $O(W)$, respectively, where N is the number of prefixes and W is the longest prefix length.

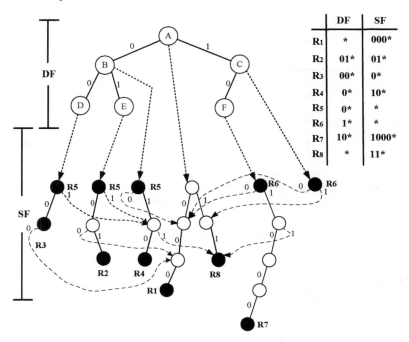

	DF	SF
R₁	*	000*
R₂	01*	01*
R₃	00*	0*
R₄	0*	10*
R₅	0*	*
R₆	1*	*
R₇	10*	1000*
R₈	*	11*

FIGURE 4.4
Grid of tries of two fields.

4.2.3 Grid of Tries

The Grid of Tries scheme reduces the amount of memory needed to store replicated nodes while keeping the access time of the Set Pruning trie. To minimize the replication of nodes (and subtrees) as in the Set Pruning trie, pointers to original rule locations (in the DF zone) are used instead. Figure 4.4 shows the Grid of Tries approach in the example above. As the figure shows, there are additional pointers from branches in the SF zone belonging to longer prefixes in the DF zone pointed to branches of shorter DF prefixes.

As in our example, a packet with destination address 000... and source address 000... also matches R1 after seven memory accesses. The search complexity of this scheme is similar to that of the Set Pruning trie.

4.3 Geometric Schemes

The presentation of a classifier where the domain of a rule is represented as a bi-dimensional object, with one axis representing the range of number of

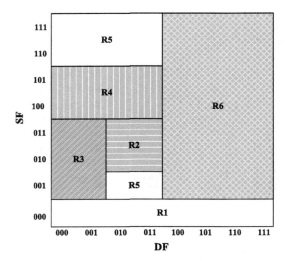
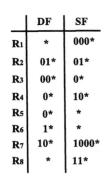

FIGURE 4.5
Geometrical representation of a two-dimension classifier.

one field and the other axis representing the range of numbers of another field, is called geometric. A geometric representation of a classifier can help to determine matching of a packet with the set of rules in the classifier. Figure 4.5 shows an example of a geometric, or more specifically a bi-dimensional representation of the classifier used in the previous tree-based examples.

As the figure shows, out of the eight rules in the classifier, only six of them, R1 to R6, are actually shown and R7 and R8 overlap over an area with high-priority rules. The absence of R7 and R8 in the bi-dimensional representation indicates that these rules are never matched by a flow. A geometric representation of a classifier is helpful in checking for rules that may not be matched, as our example shows.

4.3.1   Crossproduct Scheme

The crossproduct scheme is a geometric approach, where rules are represented in zones of intersection of two fields [163]. The crossproduct approach may be applied to classifiers with multiple fields.

As the schemes discussed show, a match occurs when a packet's metadata, at the different layers, match a rule. This matching must occur in all the fields of the rule, simultaneously. This is equivalent to a product of the values of all fields of a rule, and it is called a crossproduct.

The crossproduct scheme may consider two fields in a geometric representation, as discussed in Section 4.3. However, each product generated by each pair of values of the two different fields is assigned the rules that include those

DF	000	001	010	011	100	101	110	111
111	R5, R8	R5, R8	R5, R8	R5, R8	R6, R8	R6, R8	R6, R8	R6, R8
110	R5, R8	R5, R8	R5, R8	R5, R8	R6, R8	R6, R8	R6, R8	R6, R8
101	R4	R4	R4	R4	R6	R6	R6	R6
100	R4	R4	R4	R4	R6, R7	R6, R7	R6	R6
011	R3	R3	R2	R2	R6	R6	R6	R6
010	R3	R3	R2	R2	R6	R6	R6	R6
001	R3	R3	R5	R5	R6	R6	R6	R6
000	R1	R1	R1	R1	R1	R1	R1	R1

DF

FIGURE 4.6
Example of crossproduct table.

values. Figure 4.6 shows the crossproduct of the values of the classifier of our example.

In classifying a packet, the values of fields in the packet are matched to the axis of the geometric representation and the corresponding crossproducts indicate the matching rule. The search for all the possible values (for large ranges or the wildcard) may take a large number of memory accesses.

One well-known disadvantage of this scheme is the requirement of very large amounts of memory to represent the crossproduct. This memory issue may sever scalability for classifiers with a large number of rules.

4.3.2 Hierarchical Intelligent Cuttings (HiCuts)

Search on geometrical representation of the intersections of two fields may require multiple memory accesses, or being too slow, for classifying packet at very high data rates. An alternative strategy is to represent the geometrical representation as a tree. This tree may not necessarily be a binary one; nodes may have two or more children. One example of this representation is the hierarchical intelligent cuttings (HiCuts) scheme [76].

In HiCuts, a geometric representation (Section 4.3) may be cut into sections along an axis and the rules included in each section may be listed as being the content of a node. The selected axis may be cut in any power of two number of sections. Further cuttings may be done to each section, on the same or the perpendicular axis. A condition to stop the cuttings may be the number of rules that may be included in a node.

Using our example classifier of two fields (Figure 4.5), the geometrical representation shows an area that is 8×8 (3-bit fields). Expansion of field length can be performed to extend the area for defining it at a higher resolution. Here, DF is on the X axis and SF is on the Y axis. HiCuts does not provide guidelines on which axis to start and the number of cuttings needed. However, the goal is to obtain a representative short tree so as to speed up the classification time. Figure 4.7 shows the cuts performed to our example and the resulting tree. First, let us set as a goal to list at most two rules per node. Therefore, cutting of the areas will continue until each node holds up to two rules.

In this example, one cut is performed first, sectioning the X axis in half (Figure 4.7(a), which is denoted as (original area, axis, number of sections) or (8x8, X, 2) for this particular cut. Figure 4.7(b) shows this first cut as the root of the tree, with a node holding the label X-2, which indicates the axis and the resulting number of sections. The figure also shows a dashed line cross-cutting the classifier along the X axis. Then, there are remaining two 4x8 sections.

The right section has accomplished the goal, by including two rules, R6 and R1 (in fact it is a portion of R1), in the section (or node), but the left section (from 000 to 011 in the X axis) has more than two rules included. The root node has now a right child holding R1 and R6. The left section is now cut across the Y axis and denoted by (4x8, Y, 2) and the node holds the label Y-2. In the remainder of this description, a section of the bi-dimensional plane is represented as (range in the X axis, range in the Y axis). The section (000-011, 100-111) includes two rules, R4 and R5, but the section (000-100, 000-011) includes four rules. The cutting process in the latter section continues. Here, according to the boundary between R2 and R5, a cross-cut along the Y axis can be used to separate them. Therefore, the next cut is along the Y axis of this section (4x4, Y, 2). This last cut leaves R2 and R3 in the section (000-011, 010-011), while the section (000-100, 000-001) includes tree rules. This last section is cut again, but this time along the X axis (4x2, X, 2). This last cut leaves two sections, (000-001, 000-001) and (010-011, 000-001), including R3 and R1 in the first one, and R5 and R1 in the last one. Figure 4.7(a) shows the cuttings on the classifier and Figure 4.7(b) shows the final tree.

In the matching process, the fields of a packet are matched to DF and SF fields of the classifier. The matching is performed on the tree, starting from the root node. As before, a packet with destination address 000... and source address 000... would first match the left child of the root, and continue the way down the tree until reaching the node with R1 and R3, from which R1 is matched, as this is the highest priority rule. The number of memory accesses would be five as we consider one memory access for each node of the tree.

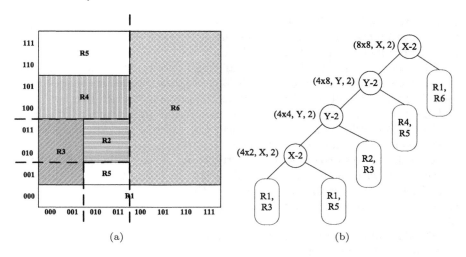

FIGURE 4.7
HiCuts example: (a) Cuts in a geometrical representation of a classifier and
(b) Tree representation

HiCuts does not include an optimized form for performing cuttings and
some decisions may be difficult to make in large classifiers. This scheme inherits the large memory needed to obtain the tree and the range matching needed
at each of the decision nodes (i.e., on what section the matching continues).

4.4   Heuristics Schemes

Here we present a very fast scheme to perform packet classification, called
recursive flow classification (RFC). However, after studying it, the reader may
associate this scheme as a geometric scheme as it shares many properties
with schemes in that category. Nevertheless, this scheme presents strategies
to simplify the search for products between fields.

4.4.1   Recursive Flow Classification

The recursive flow classification scheme aims to represent the product of fields
in an encoded format and by using partial products. It identifies different
numbers (or range of numbers) in a field and codes them with an identification
number. For instance, let us consider a range of numbers from 0 to 9. If only
numbers 5 and 6 are used from this range, 5 becomes identified as the *first*
number and number 6 as the *second* number. In this way, the number of Ids is

reduced to two. Then the numbers are linked to the rules that include them. In RFC, products between two or more fields are performed successively until the crossproduct of all fields is completed.

RFC searches multiple fields in parallel and performs the partial products in sequence. As products are sorted in phases, these phases are sequenced; a following phase requires the results from the previous phase. The last phase includes the rule number instead of the crossproduct in the content of a cross-product table, called *chunk*. The number of accesses to perform classification is equivalent to the number of phases.

The scheme uses two different data structures, one actually used for matching the fields with the information of a packet traversing the router that performs classification, or the data plane, and the other (construction plane) for calculating the values stored in the data-plane data structure, or construction plane.

The data plane includes chunks and an index calculation point (ICP) that performs the crossproduct between the contents of different chunks. A chunk stores the values of a field used in the classifier. A chunk may contain the whole header field or a portion of it; the number of bits used in a chunk may be smaller than the total number of bits of a field to reduce the number of crossproducts. For instance, instead of using a chunk for the 32 bits of an IP address, two 16-bit chunks can be used (an example of this segmentation is presented below). Consider the classifier in Table 4.2. This classifier has five rules. The 32-bit network layer (L3) addresses are segmented into two 16-bit chunks. Figure 4.8 shows the processing of chunks, their data, and the operations that ICPs apply to determine the chunk entry that is considered during the matching process. The figure shows three phases (Phases 0, 1, and 2). Phase 0 includes five chunks, which are processed in two groups, one with three chunks and the other with two chunks. The results of the crossproduct of the two groups are stored in two chunks in Phase 1. The crossproducts of the two chunks in Phase 1 are then stored in the final chunk in Phase 2.

TABLE 4.2
Example of a classifier with five rules.

Network-Layer Destination (addr/mask)	Network-Layer Source (addr/mask)	Transport-Layer Protocol	Transport-Layer Destination Port
0.94.32.0/32	*	TCP	*
0.94.2.0/32	0.0.3.2/32	TCP	range 23-88
0.94.32.0/32	*	*	80
*	*	*	80
*	0.0.3.2/32	*	*

Each chunk's entry is the Id of a different value. The classifier in Table 4.2 is partitioned as Figure 4.9 shows.

Table 4.3 shows these chunks with the actual classifier values. The selection of chunks that are crossproducted follows. Here, Chunks 0, 1, and 4 are in one crossproduct and Chunks 2, 3, and 5 are in another.

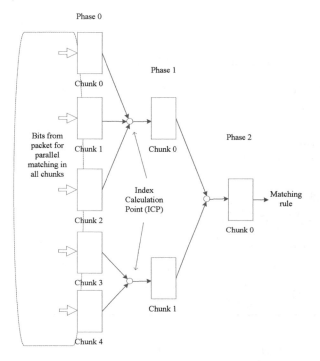

FIGURE 4.8

Example of RFC data structure for rule and packet matching (data plane).

Figure 4.10 shows the crossproduct operation of the group with Chunks 0, 1, and 4. Phase-0 chunks store the Ids of the field values corresponding to each chunk. For example, the content in Chunk 4, which corresponds to the transport layer protocol, includes the Ids of the different transport protocols in the classifier. As the field is 8 bits long, the chunk has 256 entries, and * corresponds to entries 0 to 255, and TCP to 6. Let's assign 0 as an arbitrary Id for * entries and 1 to the TCP entry. Chunk 4 then has two different Ids, 0 and 1, as the figure shows.

The information used to build the entries of the chunks in the following phase is the equivalent class ID (eqID) and a class bitmap (CBM) field. The eqID field is the identification number used by an entry in a chunk and the CBM indicates which rule includes the eqID. This table is used to calculate the content of a chunk. In Chunk 4 of our example, eqIDs are 0 and 1 as there are only two different protocol values. The CBM of the chunk for eqID 0 (or *) is 00111 as * (0-255) appears in Rules 2, 3, and 4. Therefore, eqID is represented as the bit string $b1, \ldots, bn$ for a classifier with n rules. Similarly, the CBM of eqID 1 indicates that TCP appears in all the rules, as 6 is included

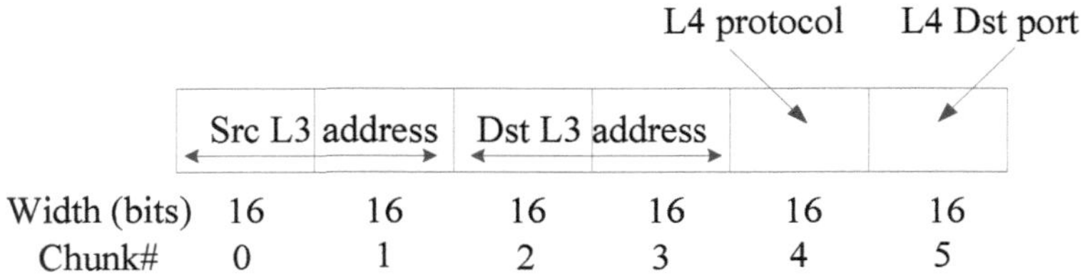

L4 protocol   L4 Dst port

Src L3 address	Dst L3 address		

Width (bits)	16	16	16	16	16	16
Chunk#	0	1	2	3	4	5

FIGURE 4.9
Segmenting of fields of classifier in Table 4.2 into chunks.

TABLE 4.3
Classifier in Table 4.2 with partitioned L3 address fields.

Rule	Chunk0 (Src L3 bits 31...16)	Chunk1 (Src L3 bits 15...0)	Chunk2 (Dst L3 bits 31...16)	Chunk3 (Dst L3 bits 15...0)	Chunk4 (L4 protocol) [8] bits	Chunk5 (Dstn L4) [16] bits	Action
(0)	0.94/16	32.0/16	0.0/0	0.0/0	TCP (6)	*	permit
(1)	0.94/16	2.0/16	0.0/16	3.2/16	TCP	range 23-88	permit
(2)	0.94/16	32.0/16	0.0/0	0.0/0	*	80	permit
(3)	0.0/0	0.0/0	0.0/0	0.0/0	*	80	deny
(4)	0.0/0	0.0/0	0.0/16	3.2/16	*	*	permit

in the 0-255 range (the packet that would match TCP as protocol would also match the wildcard *).

The values of a chunk entry and CBM of the following phases are slightly more complex, but they follow the same principle as those in Phase 0. Let's consider Chunk 0 of Phase 1. This chunk now has 12 entries as the number of eqIDs from the factor chunks (Chunks 0, 1, and 4 in Phase 0) are two, three, and two, whose product is 12. The ICP is then used to determine the entry index. The content of an entry of a chunk is identified as c indexed by the phase and chunk numbers. Therefore, c10 is the content of Chunk 0 in Phase 1. The index of each entry in c10 is index=c00*6+c01*2+c04. The first entry (index = 0) in c10 is the intersection of the CBMs corresponding to the content from the chunks in the previous phase, or Phase 0. Table 4.4 shows these contents and their intersection (or AND product). However, instead of storing the actual AND product, the eqID of it is used. The AND product in the first entry, 00011, becomes eqID 0, the AND product in entry 7, 00111, becomes eqID 1, and the AND product 01011 becomes eqID 2.

The content of the last chunk (not depicted here) may include the rule number, which is obtained from the chunk's CBM, instead of storing an *eqID*. In this case, the rule number is the highest priority rule from those matching the *eqID* as indicated in the CBM. This may be indicated as a value between 0 and 4 (or Rule 0 to Rule 4).

In the classification of a packet, the corresponding fields of the packet are matched to the content of the chunks. As an example, let us consider that

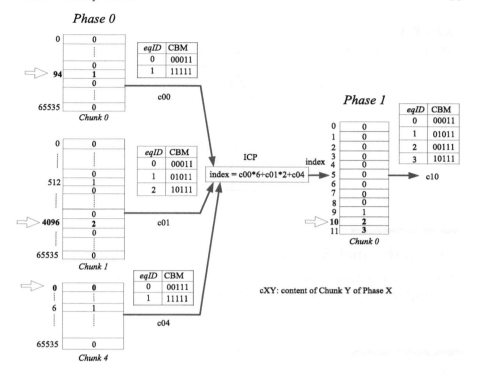

FIGURE 4.10
Portion of the implementation of RFC for the classifier in Table 4.2.

a UDP packet with destination IP address 0.94.32.0 is classified. In Phase 0, the most significant 16 bits of the address are used for matching in Chunk 0, the least significant 16 bits of the address are used for matching in Chunk 1, and the protocol field is used for matching in Chunk 4. These values will then match entries 94, 4096, and 0 in Chunks 0, 1, and 4 (as indicated by arrows in the figure), respectively, and the ICP operation matches the packet to the entry with index 10 in Chunk 0 of Phase 1, pointing to Rules 3 and 4.

As the algorithm shows, RFC reduces the number of crossproducts by partitioning rule fields. But the processing and the length of many chunks may still be large. So, while the matching speed of RFC is high (low matching complexity), the storage complexity is high, so that scalability may be compromised.

TABLE 4.4

Calculation of c10 in Phase 1.

c00	c01	c04	c00∩c01∩c03	c10
00011	00011	00111	00011	0
00011	00011	11111	00011	0
00011	01011	00111	00011	0
00011	01011	11111	00011	0
00011	10111	00111	00011	0
00011	10111	11111	00011	0
11111	00011	00111	00011	0
11111	00011	11111	00011	0
11111	01011	00111	00011	0
11111	01011	11111	01011	1
11111	10111	00111	00111	2
11111	10111	11111	10111	3

4.5   Additional Readings

There are quite a large number of schemes for packet classification. Additional schemes can be found in several surveys [10, 28, 77].

4.6   Exercises

1. Build the hierarchical tree of the two-field classifier in Table 4.5:

TABLE 4.5

Exercise classifier 1.

Rule	F1	F2
R1	01*	00*
R2	*	1*
R3	1*	110*
R4	11*	11*
R5	0*	01*
R6	*	0*
R7	11*	00*

2. Build the set-pruning trie of the classifier in Table 4.5. What is the largest number of memory accesses in this trie?

3. Build the grid of tries of the classifier in Table 4.5. What is the largest number of memory accesses in this trie?

4. Build the geometric representation of classifier in Table 4.5.

5. Build the HiCuts trie of the classifier in Table 4.5. What is the number of memory access in the resulting trie?

6. Build the Set Pruning trie of the classifier in Table 4.6.

TABLE 4.6

Classifier with seven rules.

Rule	F1	F2
R1	1*	111*
R2	00*	0*
R3	*	1*
R4	*	01*
R5	11*	*
R6	00*	1*
R7	*	00*

7. Show with a geometric representation whether or not all rules may be matched by a packet.

8. Build a HiCut trie from the classifier in Table 4.6.

9. Build the RFC data structure of the classifier in Table 4.5.

10. Build the RFC data structure for the example classifier shown in Table 4.6. Mark the path of a packet followed to match a packet with destination address 001... and source address 101....

11. Show Chunks 3, 4, and 6 of Phase 0 of the example in Section 4.4.1.

5

Basics of Packet Switching

CONTENTS

This chapter presents basic concepts, terms, definitions, and features that are commonly used in the discussion of packet switching and switch architectures. The chapter also provides an overview of the metrics used to analyze the performance of packet switches, including an introduction of basic traffic patterns usually considered in performance studies.

5.1 Introduction

Most of the information that is transmitted today through communication devices follows a packet-switching paradigm, where data are digitized in what we know as bits. These bits, taking values of 0s and 1s, are grouped into what is called a packet. Before packet switching, information would be transmitted as a continuous flow of analogous information and through a reserved path, as in the original telephone network. That paradigm is also known as circuit switching. The path followed by information would be directed according to how the circuits would get interconnected. With the emergence of packet switching, such as the Internet, discrete data could be sent through shared links in the form of packets, requiring a new set of equipment that could function in such a paradigm. This set of equipment includes what are called switches and routers. This chapter presents an overview of circuit and packet switching, and introduces major concepts used in packet switching.

5.2 Circuit Switching

The prior analog telephone network is a good example of a circuit-oriented network. This type of network requires reserving an end-to-end path before transmitting any data. This circuit cannot be shared but by the two communicating ends to transmit data. In general, a circuit-oriented network performs two functions: a) Transmission of control information, used to set the end-to-end path, through a so-called control plane. The control plane has control systems at the communicating ends that communicate to the central office. The central office determines the availability of the path and the links through which the communication will take place; b) transmission of data, which is the communication exchange performed by the two communicating ends. Information is sent as a continuous stream. To set the path, signaling takes place through the control plane. Once the path is selected and enabled, data start to flow in both directions if required (as a half- or full-duplex channel). The path used by a circuit-oriented network may be costly, as links are dedicated to the two end users who reserved the line.

5.3   Packet Switching

Packet switching uses a different working principle from circuit-oriented networks. In packet switching information is sent in discrete amounts, called packets. These packets also carry control information. It is easier to imagine the concept of a packet transmission under digitized information (as series of bits, with 1s and 0s). A packet (of bits) has boundaries, beginning and end, clearly marked. Differently from a circuit-oriented network, links of a packet-oriented network are shared by multiple users (or connections). As packets are assigned to different connections between different sets of senders and receivers, packets also carry control information, in addition to the actual data, to identify which connection the packet belongs to. These packets carrying control information are called datagrams. The control information is used to navigate through the network towards the destination. This approach means data and control paths are combined. The Internet is a prevailing example of a packet network that uses these principles.

Internet packets have variable lengths, in number of bytes. For example, speech transmitted through the Internet is partitioned into packets that may be spaced by a few other packets transmitted in-between. The transmission and conversion of voice must be fast enough so that the spacing and any underlying processes may be unnoticeable for the users performing the conversation.

The building blocks that make a packet network forward packets are switches or routers. These two blocks receive packets in one input, or line card, and forward them to an output, towards the packets' destination. Having control information (metadata) being transmitted in packets permits the network to manage scenarios of congestion or high demand of network resources to enable alternatives for making the packets reach their destinations. In this definition of switching, a switch has already "learned" the most suitable port that brings a packet closer to the destination. This transmitted control

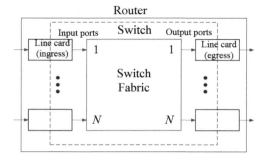

FIGURE 5.1
Top-level components of a router and switch.

information in packets may make the transmission of users' data reliable and efficient.

5.4   Routers and Switches

There are different types of equipment that are used to build a network. Here, we focus mainly on switches and routers. In general, routers operate at the network layer of the protocol stack (whether in the Open System Interconnection, from the International Open Systems [ISO], or the Internet model, from the Internet Society). This layer, which is built upon the concept of a network, uses routing to find out the path packets must follow through the network(s), from source nodes to destination nodes.

Routers have control and data planes. The control plane is in charge of finding out the ports where networks are interconnected or those that provide the most suitable path to reach a destination according to a processing policy. This information is collected through routing. The Internet was originally designed to work in a distributive manner, where routers exchange information with neighbor routers to perform topology discovery and to learn about reachable networks. The data plane is that used to forward the actual packets. Switching is a function of the data plane. Moreover, routers may use the data plane to communicate control-plane information.

Figure 5.1 shows a simplified view of the components of a router. The router has line cards with ingress and egress ports, and a switch fabric. The line cards accommodate switching (inside the dotted line) and routing (outside the dotted line) functions. As the figure shows, a router performs switching as part of its functions.

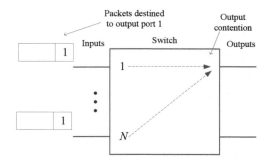

FIGURE 5.2
Example of output contention.

5.4.1 Basic Terms

5.4.1.1 Output Contention

Output contention occurs when two or more packets arrive in a switch at the same time and these are destined to the same output port. These packets must be forwarded to the output but only one can be transmitted at a time. The occurrence of output contention depends on the properties of the switching traffic. Figure 5.2 shows an example of output contention in a switch with N inputs and outputs. The figure shows two packets, one at input 1 and the other at input N, going to output port 1. These two packets would then contend for access to that output.

5.4.1.2 Blocking and Nonblocking Fabrics

Switch fabrics where packets arriving in different inputs and destined to different available outputs may be blocked from being forwarded to their outputs because of internal contention are called blocking. A switch with a blocking fabric is then called a blocking switch. Blocking switches may have a limited switching performance. On the other hand, a switch fabric that is not blocking is called nonblocking.

A fully interconnected switch fabric is an example of a nonblocking fabric as there is no internal contention for any combination of packets destined to different outputs. Figure 5.3 shows a fully interconnected fabric. Here, any input can be interconnected with any output. Yet, nonblocking switches, as any other switches, are subject to output contention.

Figure 5.4 shows an example of a blocking switch: a Clos network. In this example, the Clos-network switch comprises three stages built with small switches interconnected to build a larger switch. Here, Inputs 1, 3, and 6 are interconnected to Outputs 8, 4, and 2, respectively. With this configuration, the switch cannot interconnect Input 2 (marked with a circle) to Output 1, both of which are idle. The desired interconnection by Input 2 may contend

Inputs   Outputs

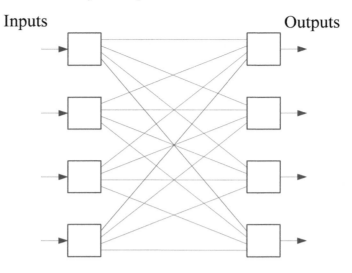

FIGURE 5.3
A fully interconnected network.

for internal links, and therefore, the switch may be blocking. Note that this example has no output contention. Yet, the switch cannot accommodate the desired interconnection. A change of configuration may, however, be able to accommodate the interconnection.

5.4.2   Classification of Switches

Switches can be classified by the number of stages, number of paths from one input to one output, whether they are blocking or not, according to the queueing strategy, and whether they perform time or space switching.

5.4.2.1   Number of Stages

The number of stages of a switch may be considered as the number of internal switch elements a packet needs to traverse to reach the destined output. A fully interconnected network is a single stage switch as a packet has to pass through the switch element at the output (the input switch could just broadcast the packet). However, a fully interconnected network has a large number of lines. A practical example is a crossbar switch. In this switch, an input may get interconnected with any other output.

Figure 5.5 shows an example of a crossbar. This fabric is the most popular as it is equivalent to a fully interconnected network, and therefore it is nonblocking. A switch fabric has N^2 crosspoint switches (also called switch elements). The figure shows the conceptual crossbar with crosspoint switches

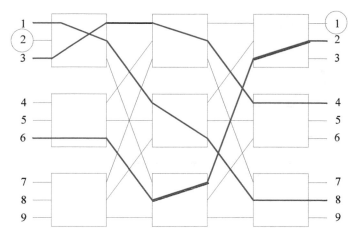

FIGURE 5.4
Example of blocking in a Clos-network switch.

(squares) in gray color and the actual connections in a crossbar in black color. A crossbar switch can be seen as an interconnected grid of crosspoint switches, as the figure shows. The horizontal lines incoming to the switch are the inputs and the vertical lines, at the bottom, leaving the switch are the outputs. An input may be interconnected with an output through a crosspoint switch. It should be noted that when a crosspoint switch interconnects an input with an output, the crosspoint activates a gate (or a multiplexer) that interconnects that input with the output, and the packet does not need to traverse another crosspoint switch. Each output also has a multiplexer controlled by an output arbiter so as to allow a single input to be interconnected to the output. The output arbiter activates the interconnection through the crosspoint control block.

The figure also shows two examples of interconnections between two inputs and two outputs, where Input 2 sends a packet to Output 1, and Input 3 sends a packet to Output 4. The black lines show the paths followed by the packets. The figure shows that only one crosspoint switch is activated per interconnection, and there is no internal contention for links.

A common property of multistage switches is that they may use a smaller number of switch elements than single-stage switches, but sometimes at the cost of being blocking. Banyan and Clos-network switches are examples of multistage switches. Figure 5.6 shows an example of an 8×8 Banyan switch. The interconnection network of the switch has three stages (packets traversing the switch must pass through at least three switches). Each switch element in a Banyan network is a 2×2 switch. A Banyan network uses different bits of the packet destination to route the packet through the different stages of the network. This switch is blocking as packets destined to different output may

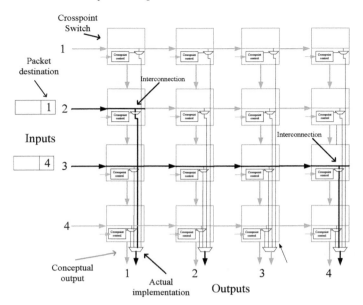

FIGURE 5.5
Example of a 4×4 crossbar switch fabric.

contend for internal links. Blocking in a Banyan network can be removed if inputs with packets are contiguous and if the packets are sorted per destination (in either ascending or descending order). There are several networks that fall in the Banyan category, such as delta and omega networks.

This figure also shows an example of blocking in the Banyan network. Here, Inputs 1 and 3 each have a packet destined for Outputs 3 and 2, respectively. Although the packets are destined to different outputs, they contend for the lower link of a switch element in the middle stage, as the figure shows. This means that these two packets cannot be transmitted to their outputs at the same time, even though when they come from different inputs and are destined for different outputs (there is no output contention). Note that in a Banyan network, there is a single path from any input to any output.

A Clos-network switch is also a multistage switch. This network generally has three stages. Different from a Banyan and crossbar switches, the Clos network uses small $n \times m$ switches, where n and m are the number of inputs and outputs of the switches, respectively. This switch provides multiple paths to interconnect an input to any output, so this is a multipath switch. The Clos-network switch is used to build switches with a very large number of ports. As the figure shows, a Clos-network switch may also be blocking, as two packets going to different destinations may contend for an internal link.

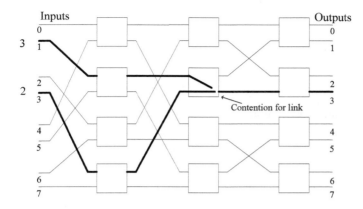

FIGURE 5.6
Example of an 8×8 Banyan switch fabric.

5.4.2.2 Number of Paths

Switches may have a single path to interconnect an input to an output. Examples of these switches are a fully interconnected network, crossbar, and Banyan networks. A Clos-network switch provides multiple paths from an input to a given output. Other examples of multipath switches are those with multiple planes, where each plane transmits a packet from an input to an output.

5.4.2.3 Queueing Strategies

Packets that cannot be transmitted to its destined output to avoid output contention are queued in the switch. The queueing may occur at the inputs (input queueing), outputs (output queueing), in the fabric (internal queueing), or a combination of these in a switch. These queueing strategies use dedicated queues. Figure 5.7 shows some examples of these queueing strategies. A combination of these queueing strategies may be applied. For example, Figure 5.7(d) shows a combination of input and output queueing.

In contrast, there are switches where the queues are shared, such as a shared-memory switch. In a shared-memory switch, inputs and output may all share the same memory, but each may have a logical queue. These switches also share access time among ports. Ports take turns to access the memory, so these switches are categorized as time-division switches.

5.4.3 Time and Space Switching

Early packet switches have a single bus to exchange packets. So, they would segment a time slot in N small periods (i.e., mini slots) and a cell would be transmitted during one of those mini slots. To transfer such number of bits in $1/N$th of the time, the bus would be running N times faster than the

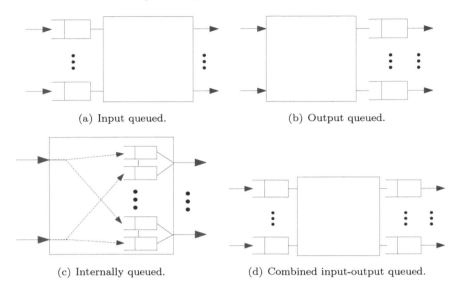

(a) Input queued. (b) Output queued.

(c) Internally queued. (d) Combined input-output queued.

FIGURE 5.7
Some queueing strategies in switches.

external links. This is what is called time switching. In contrast with more recent packet switches, each input and output port may have dedicated lines so that these lines would run at the same speed as the external links. Making use of parallel internal paths is called space switching.

5.4.4 Managing Internet Packets in a Switch

5.4.4.1 Segmentation and Reassembly

The length of Internet packets is variable. The Internet Protocol version 4 (IPv4) uses a header field to indicate the packet length. In IPv4, the smallest packet length is 20 bytes long (a header with no options and no payload) and the larger packet length is practically defined by the maximum transmission unit (MTU), as stipulated by the data-link layer.

It is expected then that switches handle variable-length packets internally. However, early work on fast packet switching considered Asynchronous Transfer Mode (ATM) as the technology of choice. ATM adopted the use of fixed-size packets, called cells, and the design of the many ATM switches proposed were, therefore, also cell based [135]. A switch for IP traffic may then segment variable-length IP packets when they arrive in the switch, into fixed-size portions of a packet, called cells for internal handling. Packets are then reassembled into their original variable-size before departing the switch. These cells, however, are not ATM cells; they are fixed-length segments of an IP packet. The remainder of this chapter, and most chapters of this book, refer to fixed-length packets or cells.

One of the advantages of switching packets in the form of cells is that the design of switches and the management of packets may be simpler than using variable-length packets. For example, the transmission time of a variable-size packet depends on the length of each packet. But in fixed-size cells, the transmission time is constant, $\frac{L}{R}$, where L is the cell length and R is the transmission speed of the external links. A constant transmission time is called time slot. L can be selected by the switch designer according to the desired properties of the switch [5]. For example, a large L may reduce the required speed in which a switch may be re-configured (as the time slot is longer) but for traffic carrying mostly small IP packets, more padding may be required. A large amount of padding may waste bandwidth and energy. On the contrary, if a small L is chosen, less padding may be required by the switch but the switch may have a shorter time for reconfiguration; that would require fast management of the switch. These cells can be switched to the outputs in cell- or packet-based modes. these modes are described as follows.

5.4.4.2   Cell-Based Forwarding

In cell-based forwarding, decisions on which cell is switched to the output at a given time slot are performed independently of which cell was selected in the previous time slot. Re-assembly of packets at the output ports may require multiple queues; one per packet, to wait for cells of each packet for re-assembling. It is desirable that cells arrive in order —the first part of the packet first and the last part of the packet at the end— to avoid increasing the complexity of the re-assembling mechanism. The number of queues may be in the order of the number of packets being interleaved during the switching time.

5.4.4.3   Packet-Based Forwarding

In packet-based forwarding, once the first cell of a packet is switched from a given input to the destined output, the output is reserved to continue receiving the remainder cells of that packet. That means that the input and output may remain switching cells consecutively until the last cell of the packet is forwarded. This input-output path of the switch may not be reconfigured during the time the cells of the packet are forwarded.

Figure 5.8 shows an example of these two forwarding modes. Figure 5.8(a) shows packets A and B arrive in different inputs of a switch but they are destined to the same output. Figure 5.8(b) shows possible forwarding order of cells in packet- and cell-based forwarding modes. In cell-based forwarding, cells from packets A and B are interleaved (as other inputs may not have packets for the same output) and in packet-based forwarding, all cells from each packet are forwarded consecutively.

Each mode has advantages and disadvantages. The designer must select

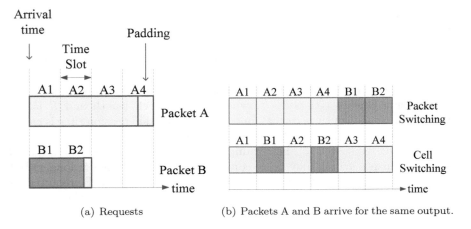

(a) Requests   (b) Packets A and B arrive for the same output.

FIGURE 5.8
Cell- and packet-based forwarding.

which of these modes provides the sought benefits. Using packet concatenation in cell switching may increase the utilization of link bandwidth [5]. Packet concatenation is a hybrid mode that combines cell- and packet-based switching modes. In packet concatenation, cells from the same flow may be used to pad incomplete cells (after segmentation).

5.5   Performance Metrics

There are several metrics used to define the features of packet switches and routers.

5.5.1   Throughput

Throughput may be defined in different terms. The throughput of a switch port may be indicated as the number of bits (or packets) per second as an absolute measure, as T_l. In this case, the maximum throughput can be considered as the port or link capacity, C_l, and $T_l = C_l$.

The normalized throughput, T_n, can be represented as the number of bits per second of the outgoing traffic of a device (e.g., router or switch) over the number of bits per second of the incoming traffic.

$$T_n = \frac{\text{Number of outgoing bits per second}}{\text{Number of incoming bits per second}}$$

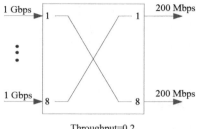

FIGURE 5.9
Example of throughput and normalized throughput.

$$T_n = \frac{\sum f_o}{\sum f_i} \qquad (5.1)$$

where f_o and f_i are the number of bits per second of the outgoing and incoming flows, respectively.

The normalized throughput can also be obtained from rates represented in packets per second if packets have a fixed size:

$$T_n = \frac{\text{Number of leaving packets per second}}{\text{Number of incoming packets per second}}$$

The normalized throughput is used as a practical indicator of the performance of a packet switch or a router as it is not limited to an absolute number or bits per second. For example, consider the black box device in Figure 5.9, which has eight ingress and egress links, each with $C_l = 1$ Gbps. If the black box switches a traffic amount of up to 20% of the port capacity, $T_n = 0.2$. Should we be interested in the absolute numbers, one could find out that $T_l = 200$ Mbps. Therefore, the normalized throughput is a general and convenient representation.

5.5.2 Latency

The time packets spend in a switch may be caused by queueing and processing (response) times. These two general terms are described as follows.

5.5.2.1 Queueing Delay

Packets passing through a switch with multiple inputs and outputs may spend some time inside before they depart from it. This latency may be caused by two events: the time it takes to process a packet, such as the response time that an interface may take to read or process part of the packet header, and the time the packet spends making a queue in the switch. While the first one is a constant delay and it may not affect the performance of the switch if the delay is small, the queueing delay may vary with time, according to the amount and distribution of the incoming traffic load and the policies and switch architecture.

5.5.2.2   Processing Delay

The response time depends on design, physical properties of the device, and the functions associated with it. For example, if address lookup is included, the time to process lookup is also added into the response time. In another example, if the processing of data involves accessing memory, the time to process the data will include the memory access time.

5.5.3   Internal Transmission Speed and Bit Slicing

The switch fabric of a router may use same or different transmission speed as the external links. The external links of a switch are considered to transfer bits in serial, one bit at a time. As transmission speed increases, multiple lines may be used as a combination of serial and parallel transmissions. The internals of a switch may transfer data with a different level of parallelism. For example, a packet or cell can be transmitted from an input to an output of a switch in the shortest time if L bits are transferred in a single clock cycle. The duration of the clock cycle used for receiving (or transmitting) a cell to the external lines are used as reference. In the external lines of a switch, the duration of a clock cycle is the time it takes to transmit one bit at the line rate.[1] The transmission of L parallel bits would require a serial-parallel conversion at the input ports and a parallel-serial conversion at the output ports.

The internal circuits of a switch may be slowed down or sped up by transmitting multiple bits at a time while using a different internal (clock) rate than that used by the transmission links. Multiple bits may be transmitted in parallel to what it may appear to avail multiple switching planes. An example of a switch based on bit slicing is the Tiny-Tera [112].

For example, an L-bit long cell, arriving at $\frac{L}{t}$ bits per second (bps), where t is the period of the clock cycle of the line rate, could be transmitted to the output in a pair of bits at a time during the time period t, or at $\frac{2L}{2}$ bps.

5.5.4   Speedup

Speedup is the ratio between the speed in which data is transmitted inside a switch and the data rate of the external links. A speedup equal to one means

---

[1]Fast transmission technologies actually send parallel bits through multiple lines, as a combination of serial-parallel transmissions.

that the internals of the switch and the link rate run at the same speed, or have no speedup.

Speedup may be used when the width of a bus (in a shared-bus system) may transmit more than one packet within the time a packet is received. The bus may also run at higher speed to keep up with the external link speeds. Speedup may be used in packet switching to increase the performance of a switch [38, 45, 150]. But this speedup is only possible if all components of the switch are able to keep up with that speedup. Memory is a component that often cannot be sped up.

Speedup can be achieved by transmitting multiple bits or packets in parallel running at the same rate as the external lines (space-based speedup) [23], by using a faster clock in the switch, which would also switch bits at a faster rate than the external lines (time-based speedup), or a combination of both.

5.5.5 Unicasting and Multicasting

Networks may support not only unicast but also multicast services. Routers and switches that support the network services run functions that do so: routers execute multicasting protocols (multicast routing and Internet Grouping Management Protocol (IGMP) [24] among others) and switches must be able to replicate multicast packets if the switch branches out a multicast route, so packets can be sent towards their multicast destinations.

5.5.5.1 Unicasting

A unicast packet is the one sent to a single destination. In this way, this packet represents a one-to-one communication service. Switches are primarily designed to provide unicast service.

5.5.5.2 Multicasting

A multicast packet is directed to one or multiple destinations. Switches supporting multicasting must be able to replicate a multicast packet for as many destinations. The capability of providing multicast services may impact how the switch works and its performance. In general, the interaction between unicast and multicast packets needs to be defined before the switch is designed.

In a switch, a multicast packet that is sent through multiple output ports must be efficiently replicated. Management of multicast packets may require queueing, scheduling, and recordkeeping. Replication of packets may be performed at arrival time, or at the forwarding time. Either choice affects the complexity of the switch.

5.5.5.3   Multicast Queueing

Multicast packets can be queued at the inputs of the switch in queues dedicated to multicast traffic. These queues may store a multicast packets. Replicated multicast packets may also be stored as unicast packets. This means that after a packet is replicated, it is sent to the queue corresponding to the destined output port.

5.5.5.4   Multicast Scheduling

Multicast packets may need to be replicated and forwarded by a switch. Different scheduling policies may be applied. These are call splitting policies [33, 82]. Some of these policies are strict-sense call splitting (SSCS), wide-sense call splitting (WSCP), one shot, and revision scheduling. Figure 5.10(a) shows an example of multicast traffic to be switched to their destined output ports. In Figures 5.10(b)-5.10(e), ϕ indicates the selected request from an input (row) to an output (column). In one shot scheduling, all requests from the same row (input) are selected (Figure 5.10(b)). In SSCS, at most one request is selected for one row (Figure 5.10(c)). In WSCP, there is no restriction on how many requests are selected per row (Figure 5.10(d)). In revision scheduling, one-shot scheduling and WSCP are combined (Figure 5.10(e)).

$$
\begin{bmatrix} 1 & 1 & 0 & 0 & 1 \\ 0 & 1 & 0 & 0 & 0 \\ 0 & 0 & 0 & 1 & 0 \\ 0 & 1 & 1 & 0 & 0 \\ 0 & 0 & 1 & 1 & 0 \end{bmatrix}
\begin{bmatrix} \phi & \phi & 0 & 0 & \phi \\ 0 & 1 & 0 & 0 & 0 \\ 0 & 0 & 0 & \phi & 0 \\ 0 & 1 & 1 & 0 & 0 \\ 0 & 0 & 1 & 1 & 0 \end{bmatrix}
\begin{bmatrix} \phi & 1 & 0 & 0 & 1 \\ 0 & \phi & 0 & 0 & 0 \\ 0 & 0 & 0 & \phi & 0 \\ 0 & 1 & \phi & 0 & 0 \\ 0 & 0 & 1 & 1 & 0 \end{bmatrix}
\begin{bmatrix} \phi & 1 & 0 & 0 & \phi \\ 0 & \phi & 0 & 0 & 0 \\ 0 & 0 & 0 & \phi & 0 \\ 0 & 1 & \phi & 0 & 0 \\ 0 & 0 & 1 & 1 & 0 \end{bmatrix}
\begin{bmatrix} \phi & \phi & 0 & 0 & \phi \\ 0 & 1 & 0 & 0 & 0 \\ 0 & 0 & 0 & \phi & 0 \\ 0 & 1 & 1 & 0 & 0 \\ 0 & 0 & \phi & 1 & 0 \end{bmatrix}
$$

(a) Requests   (b) One shot   (c) Strict sense   (d) Wide sense   (e) Revision

FIGURE 5.10
Call-splitting strategies.

These scheduling policies have implications in the design of the switch. In general, one-shot may be easy to manage but performance may be affected. On the other hand, WSCP may allow to achieve higher performance but it may be complex to manage. The management complexity may also affect the overall performance of the switch and the cost of the switch.

5.6   Traffic Patterns

The throughput or any other performance parameter that a packet switch may achieve may be specific to a traffic pattern. That is, a packet switch may achieve certain performance for specific traffic patterns. If a switch can handle any traffic pattern that the switch may receive, it is said that the switch is able to perform under admissible traffic. This traffic is defined as follows.

5.6.1 Admissible Traffic

Admissible traffic is the one that does not overbook the inputs and outputs of a switch; that is,

$$\sum_i \lambda_{i,j} < 1.0$$

and

$$\sum_i \lambda_{i,j} < 1.0 \tag{5.2}$$

where $\lambda_{i,j}$ is the data rate from input i and going to output j of the switch. These inequalities indicate that traffic coming to an input for any output may not be more than the capacity of the input and traffic going to any output may not be more than the capacity of the output, respectively. The value 1.0 is the normalized throughput of an input or output. A packet switch that is overbooked is not expected to handle beyond the capacity of the switch. Yet, overbooking traffic patterns may be used to test whether a switch or a port can reach full utilization of port capacity.

The performance of packet switches (or switching function of routers) may be tested under specific traffic patterns, experimental traffic, or traces. Traffic characteristics may be classified into two different categories: 1) arrival pattern and 2) destination distribution.

5.6.2 Arrival Distributions

Packet arrivals may be coarsely classified into individual and bursty. Individual arrivals describe traffic where single packets may be transmitted from source to destinations at random times. The source may generate a number of packets, consecutively or sporadically, but a common feature is that packets are independently generated. This traffic model is called Bernoulli arrivals.

Traffic can also be generated in the form of trains of packets, where it is most likely that once the train has started (or stopped), the state will remain. This traffic is called bursty and is often described as On-Off traffic as a two-state modulated process.

5.6.2.1 Bernoulli and Poisson Arrivals

Let k be the number of successes in n independent Bernoulli trials, then the probability of k successes, $P_n(k)$, is given by the binomial probability:

$$P_n(k) = \binom{n}{k} p^k (1-p)^{n-k} \tag{5.3}$$

where the binomial coefficient

$$\binom{n}{k} = \frac{n!}{k!(n-k)!} \tag{5.4}$$

For a very large n,

$$p_k \simeq \frac{\alpha^k}{k!} e^{-\alpha} \tag{5.5}$$

where $\alpha = np$.

5.6.2.2   Bursty On-Off Traffic

Bursty traffic occurs when a packet (or cell) may have higher probability of arrival if another packet has already arrived or vice versa. Bursty traffic seems to arrive in trains of cells (time slots), or On-state, or a trains of silent periods, or Off-state. This arrival pattern can be described more formally with probabilities, as to follow the state transitions Figure 5.11 shows. The probability that the On period lasts i time slots is:

$$Pr\{X = i\} = p(1-p)^{i-1}, \tag{5.6}$$

for $i \geq 1$. The average burst length, β, is:

$$\beta = E[X] = \sum_{i=1}^{\infty} iPr\{X = i\} = \frac{1}{p} \tag{5.7}$$

The probability that the idle period lasts for j time slots is:

$$Pr\{Y = j\} = q(1-q)^{j-1} \tag{5.8}$$

for $j \geq 0$.

The average idle period is:

$$\alpha = E[Y] = \sum_{j=0}^{\infty} j \, Pr\{Y = j\} = \frac{1-q}{q}$$

and the offered load ρ, or duty cycle, is the portion of a period that the state of arrivals is On (or active):

$$\rho = \frac{\frac{1}{p}}{\frac{1}{p} + \sum_{j=0}^{\infty} jq(1-q)^j} = \frac{q}{q+p-pq} \tag{5.9}$$

5.6.3   Destination Distributions

The distribution of traffic may be considered to observe how a switch handles it. Some switches may show some design issues as they may not be able to achieve high throughput or low queueing delay under specific traffic patterns.

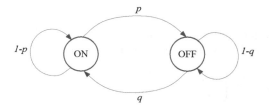

FIGURE 5.11
State diagram of the On-Off traffic model.

5.6.3.1 Independent and Identically Distributed Traffic

When studying the performance of a switch, one can consider that the traffic that arrives in an input is independent from how arrivals in other inputs occur (independent), yet the distribution of the incoming traffic at each input is similar that of the other inputs (identical). Such a condition is called independent and identically distributed (i.i.d.) traffic. The following models are examples of i.i.d. traffic, except for the example of independent and nonuniformly distributed traffic.

5.6.3.2 Uniform Distribution

It is generally considered that the traffic most benign and easy to switch is that with a uniform distribution among all outputs. Such traffic has an average rate for each output $\lambda_{s,d} = \frac{1}{N}$, where s is the source port and d is the destination port. That is, a packet that arrives in an input has equal probability to be destined to any output.

5.6.3.3 Nonuniform Distribution

There are an unlimited number of nonuniform traffic models. Here we describe several of these models.

Unbalanced traffic.

The unbalanced traffic model [139] sends a portion u of traffic from input s to output d and the rest is distributed uniformly among all outputs. That is:

$$\rho_{s,d} = \begin{cases} \rho\left(u + \frac{1-u}{N}\right) & \text{if } d = s \\ \rho\frac{1-u}{N} & \text{otherwise} \end{cases} \tag{5.10}$$

where ρ is the input load at an input port. When $u = 0.0$, the traffic is uniform and when $u = 1.0$, the traffic from input s is directed to output d, where $d = s$. The destination can be modified to be $d = s + 1 \mod N$ or any other output.

Diagonal traffic.

In this traffic model, a portion g, where $0 < g < 1$, of the traffic is sent from input s to output $d = s \mod N$, and the remainder $1 - g$, to $d = s + 1 \mod N$:

$$\rho_{s,d} = \begin{cases} \rho g & \text{if } d = s \\ \rho(1 - g) & \text{for } d = (s + 1) \mod N \end{cases} \tag{5.11}$$

Hot-spot traffic.

One-shot traffic occurs when all inputs send most or all incoming traffic to a single output. In such case,

$$\rho_{s,d} = \begin{cases} \rho\left(\frac{1}{N}\right) & \text{for } d = h \\ 0 & \text{otherwise} \end{cases} \tag{5.12}$$

where h is the hot-spot output and $1 \le h \le N$.

Chang's traffic.

This traffic model can be defined as an input that only sends traffic uniformly distributed to all outputs except that with the same index as the input [26]. That is:

$$\rho_{s,d} = \begin{cases} 0 & \text{for } d = s \\ \rho\left(\frac{1}{N-1}\right) & \text{otherwise} \end{cases} \tag{5.13}$$

Power-of-two (Po2) traffic

In this traffic model, an input sends a fraction of traffic that is inversely proportional to a power of two [16]. That is:

$$\rho_{s,d} = \rho\frac{1}{2^{x+1}} \text{ for } d = (s + x) \mod N \tag{5.14}$$

where $0 \le x \le N - 1$

5.6.3.4   Independent, Nonidentical, and Nonuniformly Distributed Traffic

In this traffic model, the traffic that inputs receive may be independent from each other. There are an undefined number of distributions that can fit this model. An example of a model with these features may be described as

$$\rho_{j=\{1,\ 2\}} = \rho \sum_i \rho_{i,j} \ for \ i = 1, \ldots, \ N - 1$$

$$\rho_{j=3} = \rho \sum_i \rho_{i,j} \ for \ i = 2, 3$$

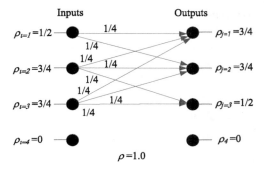

FIGURE 5.12
Example of independent, nonidentical, and nonuniform distributed traffic in
a 4x4 switch.

where

$$\rho_{i,j} = \frac{1}{N}$$

and $1 \le i, j \le N$.

Therefore,

$$\rho_{j=\{1,\ 2\}} = \rho \left(\frac{N-1}{N} \right)$$

and

$$\rho_{j=3} = \rho \left(\frac{2}{N} \right)$$

for any $N \ge 4$.

Figure 5.12 shows an example of this traffic model in a 4x4 switch. In this
example, not all the traffic received by the inputs is identically distributed.
Not all outputs receive the same load and not all inputs receive the same load.
Nevertheless, the traffic remains admissible.

5.7   Ideal Packet Switch: Output Queueing

The output-queued (OQ) switch is considered to be the ideal packet switch
as it makes arriving packets (or cells) available at their destined outputs,
without concerns for output nor internal contention, the following time slot.
This means that it achieves 100% throughput for any admissible traffic pattern
and it may provide performance guarantees as packets are available at the
outputs. These guarantees may depend on having a controlled incoming traffic
[123, 186] and on the properties of the scheme used to dispatch packets at the
output.

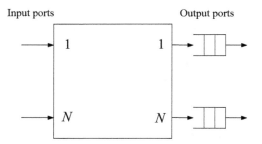

FIGURE 5.13
Output-queued switch.

In the OQ switch, cells are forwarded to their destined outputs as they arrive. If multiple cells are destined to the same output, the output queue may store the cell from the input with the smallest index number first, then the cell from the second smallest index next, and so on. To accommodate the storing of cells within the time slot, the memory (used to queue packets) and the fabric of the switch run up to N times faster than the link rate, where N is the number of input ports. This means that this switch must adopt a speedup, S, equal to N. Figure 5.13 shows an $N \times N$ OQ switch.

To show the required speedup, let's consider that each input receives one cell at a given time slot, and all arriving cells are destined to the same output port. Since the OQ switch has no queue at the inputs, N cells must be written to the output queue. At the same time, one cell may leave the switch if it has been stored in the previous time slot. That is, the queue must perform N writes and 1 read, assuming that cells are transmitted in one clock cycle. Therefore, the required speedup is $N + 1$. This is, for a memory with access time t_{mem}, the time slot may be divided in $N + 1$ mini-slots (the transmission rate is speeded up $N + 1$ times, or

$$t_{mem} \leq \frac{L}{R\,(N + 1)}$$

where L is the size of (fixed-length) cells and R is the link rate. This equation indicates that the memory access time must be smaller than or equal to the time slot divided $N + 1$ times.

However, if dual-port memories are used to implement the output queues, the read cycle can be overlapped with the write cycles. Therefore, the speedup is N, or

$$t_{mem} \leq \frac{L}{R\,N} \tag{5.15}$$

Then, as N increases, the memory speed required also increases. Considering that memory has not been keeping up with the rates in which transmission speeds grow, it becomes a limiting factor for building even moderate-size switches using this architecture. This issue has motivated a wide search for alternative architectures where speedup is minimized. These architectures are discussed in subsequent chapters. A study analysis of the OQ switch is presented in [89].

5.8 Exercises

1. Describe the advantages and disadvantages of using small or large cell sizes for segmenting variable-length IP packets for internal switching.

2. Show an example of output contention in two outputs of a 4x4 switch.

3. Modify one of the connections in Figure 5.4 to show that the three desired connections (including the one from input 2 to output 1) can be accomplished.

4. Write the relationship between the number of ports of an OQ switch and the required memory speedup if the queues are shared by pairs of ports (assume that N is even).

5. Which of the traffic models described in Section 5.6.3 may suffer from output contention.

6. What probabilities, p and q, of the On-Off traffic model may provide 60% duty period of an average six time slots in a total period of 10 time slots?

7. Solve the same question as above but for a duty period of 80% for a total period of 20 time slots.

8. Draw an 8×8 Banyan network and show the path that a cell entering in input port 011 would follow to arrive to output port 101.

9. Show an example of double HoL blocking in a 4×4 switch.

10. Show a general representation of unbalanced traffic in a 4×4 switch using a matrix representation.

11. Show a matrix representation of a 4×4 crossbar fabric interconnecting Input 1 to Output 3, Input 2 to Output 4, Input 3 to Output 2, and Input 4 to Output 1, where a matrix element 1 means cross-connection and 0 mean no cross-connection.

12. Discuss why cell switching may need larger re-assemble buffers at the outputs of a packet switch.

6

Input-Queued Switches

CONTENTS

Input-queued (IQ) packet switches are a popular switch architecture in commercial routers. A motivation to adopt them is that the required memory speed may be relaxed in comparison to other queueing strategies. This chapter discusses the challenges for this switch architecture and the matching schemes used to configure these switches.

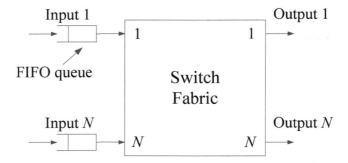

FIGURE 6.1
Input-queued switch with FIFO queues.

6.1   Introduction

Input-queued switches store packets that cannot be send to their destined outputs at the input-port queues. The queues at the inputs are used to resolve output contention. Packets that are not allowed to be forwarded to their output ports, because another packet is selected to be forwarded instead, wait at the input queue of the switch. IQ switches have the advantage of running memory at the lowest speed among most switch architectures. Therefore, these switches have greatly captured the interest of researchers and manufacturers of network equipment.

The discussion in this chapter considers that IQ switches are cell based, such that cells are transmitted in a given time slot. IQ switches require cells to be scheduled for transmission from inputs to outputs at a given time slot. The switch fabric is then configured to forward the scheduled packet at a given time. IQ switches may be built on any switch fabric. In general, we consider a crossbar as switch fabric in this chapter.

6.2   Input-Queued Switch with FIFO

The simplest method to queue packets in the inputs of an input-queued switch would be using the first-in first-out (FIFO) policy. Packets arriving in an input and destined to different output ports would be stored in the same FIFO queue. Figure 6.1 shows an input-queued packet switch with N inputs and outputs. In this switch, packets at the head of line (HoL) of the FIFO are considered for scheduling.

The input queues may write and read one cell each time slot as up to one cell may arrive and one cell may depart a switch port. Therefore, the memory access time of the memory used to build an input queue may be required to run with a speedup of two. However, for a memory with dual port, such that the write and read cycles may be performed simultaneously, no speedup is needed. The maximum memory access time t_{mem} is determined as

$$t_{mem} \leq \frac{L}{R} \qquad (6.1)$$

for cells with length L being transmitted at link rate (R).

IQ switches may require a scheduler, whether it is implemented in a centralized or distributed manner, to determine which cells are forwarded to resolve output contention. We consider that the switch fabric used by an IQ switch is nonblocking in the remainder of this chapter. The switch works as follows: Inputs notify the scheduler about the destination of the HoL cell, and the scheduler selects which input sends its HoL cell to the destined output. That is, the scheduler selects which input forwards a cell to which output. This process is also called *matching*.

IQ switches with FIFOs suffer from HoL blocking. This phenomenon is caused by the used FIFOs as they aggregate cells going to any output port. HoL blocking occurs when the HoL cell loses contention with a cell destined to the same output but from another input and the cell behind it is destined to an idle output port. In this case, the cell behind the HoL cell is blocked from being forwarded.

Figure 6.2 shows an example of HoL blocking. In this example, Inputs 1 and 4 have HoL cells destined to Output 3. Input 4 is selected to forward its cell and the HoL cell of Input 1 remains at the queue. However, Input 1 has a cell destined to Output 2, which is idle (i.e., no other input is matched to it). However, that cell is blocked by the HoL cell at the input as the queue is a FIFO.

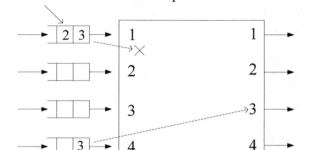

FIGURE 6.2
Example of HoL blocking.

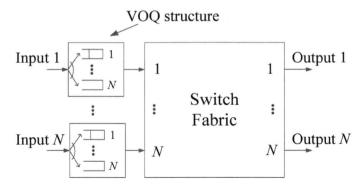

FIGURE 6.3
An IQ switch with VOQs.

It has been found that the throughput of an IQ switch with FIFOs at the inputs approaches 58.6% under uniform traffic with Bernoulli arrivals. This throughput is limited by the HoL blocking phenomenon. In a switch with FIFOs as input queues, an input can only be matched with the output for which the HoL cell is destined. Although the low complexity of these switches is appealing, their throughput is not satisfactory for building high-performance or high-capacity switches.

6.3 Virtual Output Queue (VOQ) Switches

IQ switches may use one queue to store packets destined to one output to avoid HoL blocking. These queues are called virtual output queues (VOQs). In this way, cells destined to an output port are not blocked by a HoL cell as a VOQ holds cells for the same destination only. Figure 6.3 shows an IQ switch with VOQs at the inputs. The VOQs enable a scheduler to match an input to any output of the switch for which it has a cell.

Matching is more effective in a switch with VOQs than in one with FIFOs as any input may be matched to any output. In IQ switches each input may forward one cell at a time and each output may receive one cell at a time, if there is no speedup.

With the flexibility that VOQs provide, the performance of a switch becomes a function of the matching scheme used. The performance goal of a switch is to achieve 100% throughput under any admissible traffic distribution.

To perform matching, it is necessary to list choices and preferences per each input and output. Therefore, each input may have a priority list indicating the choices for matching and the order of preference. For example, input i may list these choices and preferences as $\{k, l, m\}$. Here, k, l, and m are the possible output choices with k being the most preferred and m the least preferred one.

Priority lists may be based on the weight of each possible match to define the choice priorities or on a weightless manner. In a weighted priority list, each possible match is assigned a weight and the order of preference in the list is set by sorting the weights. The weight is determined by the policy used to make selections. For example, the weight of matching input i to output j may be the number of cells in $VOQ_{i,j}$, or the waiting time of the HoL cell in the VOQ.

In a weightless priority list, the order of the members in the list indicates the level of preference. Examples of such priority lists could be a fixed order, such as a round-robin schedule, or a dynamically changing schedule whose order is modified each time a selection is made. An example of the latter is the last recently used policy.

6.4 Weight and Size Matching

Figure 6.4 shows examples of weighted and weightless sets of match requests. In weighted matching, requests are associated with a weight and this is carried by the request. A scheduler is a set of input and output arbiters, one per input and output ports, respectively. The weights are used by the arbiters to perform selection (Figure 6.4(a)). Processing the requests in size matching may seem simpler to perform as they carry no weights (Figure 6.4(b)), but the matching complexity also depends on the dynamics of the scheme used. The match is shown as a bipartite graph, where one party is the set of inputs and the other is the set of outputs.

The objective of a match is to achieve the most effective result, which could be the highest throughput of the switch or smallest delay experienced by packets going through the switch. In a bipartite graph, each port has a priority list, in which the objective is to match a port with the counterpart at the top of the priority list. The matching process is aimed to follow the priority list. The performance of the switch depends in an extent on the metric used to establish the priority list. An example of such a process is the Galey-Shapley algorithm [78]. A weighted match may use the total sum of each individual match, as the objective of the largest weight or maximum weight matching (MWM). In contrast, a weightless or size matching has the objective of achieving the largest size match (i.e., the largest number of edges in a bipartite graph). A maximum size matching (MSM) may achieve up to N matches in an $N \times N$ switch; however, it is not clear whether MSM may achieve 100% throughput under every admissible traffic pattern.

In a match, each party selects the most preferred choice among those avail-

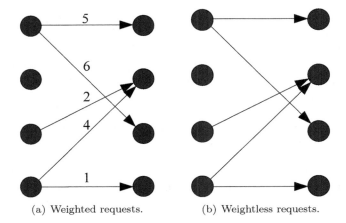

(a) Weighted requests.   (b) Weightless requests.

FIGURE 6.4
Examples of weight and weightless (size) matching requests.

able. The selection can be performed in a centralized or distributed manner. In a centralized manner, the switch may match one set of ports at a time. In distributed selection, all ports may perform matching at the same time; each port may perform the selection for a match without considering the selection made by other ports. In this case the selection may start at the output ports or at the input ports. A centralized scheduler may collect information about the cells and destinations available in each input and perform matching according to those choices. On the other hand, a distributed scheduler is partitioned into arbiters, where each arbiter is associated with a port. Arbiters collect information regarding that specific port.

Weighted matching can be roughly classified into MWM and maximal weight matching. In MWM, the resulting matches in a switch have associated a weight and the sum of all those weights is the largest possible according to the existing weights at the time of the match. In a similar way, size matching may be classified into MSM and maximal size matching schemes. We discuss MWM, and the maximal versions of weighted and size matchings in the following sections.

6.5   Maximum Weight Matching (MWM)

MWM achieves the maximum weight as the sum of the weights of all matches in a switch. One straightforward process to find an MWM is to explore all possible combinations and select one that achieves the maximum weight. The computation complexity of a scheme for this is known to be $O(N^{\frac{5}{2}})$ [81].

Figure 6.5 shows an example of a MWM match. In this example, inputs have traffic for different outputs as Figure 6.5(a) shows. Each black circle on the left side of the figure is an input and the circles on the right side are outputs. The arrows going from left to right indicate requests for a match are sent from inputs to outputs. Figure 6.5(b) shows the selected MWM from the choices presented in Figure 6.5(a). This result is the largest weight of the complete match that can be achieved in this case.

6.6   Maximal-Weight Matching

As switch fabrics have to be reconfigured each time slot. The time to perform re-configuration is the transmission time of a packet or cell, or $\frac{L}{R}$. If a cell is 64-byte long and the link rate is 10 Gbps, then the allowed reconfiguration time is 51.2 ns. For a medium-size switch, the time to perform MWM may not be long enough for high data-rate links. This is usually the case as link rates continue to increase. Despite advances in approximations with lower complexity [56], the resolution time may continue to be long. Therefore, it is practical to consider maximal-weight matching schemes as these schemes can process a matching in a fraction of the time an MWM scheme takes. However, maximal-weight matching does not guarantee an MWM, and therefore, the properties of the adopted maximal-weight matching scheme depend on its operation and selection policy.

A maximal-weight matching scheme uses the weights of the edges of a bipartite graph to decide whether an unmatched input is matched to an unmatched output. This match is usually carried without considering matches of other ports as MWM does, as long as each input is matched to one output and each output is matched to one input. That is, the selection of ports in a maximal-weight matching is performed with local information (i.e., only among those available to the port).

An example of maximal-weight match is a parallel matching scheme that uses a weighted selection policy. In this process, all input ports and output ports perform matching at the same time. Each port has an arbiter and the matching consists of three phases: request, grant, and accept. During the request phase, the inputs send requests to the ports for which they have one or more cells destined. In the grant phase, the outputs select one of the requests according to the adopted selection policy. The output arbiters send their selection to the input arbiter. In the accept phase, input arbiters select one grant per port and notify the output arbiters. This last notification confirms the match and the switch fabric is configured to transmit the HoL cells of the queue at the matched inputs. The transmission of cells may be performed after the matching process or in the following time slot.

Figure 6.5(c) shows the maximal-weight match of the requests in Figure

6.5(a). As the figure shows, the results of MWM and maximal-weight matching are different in this scenario. MWM achieves a larger summed weight than that achieved by maximal-weight matching.

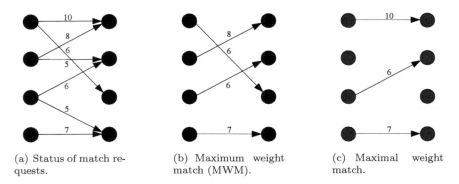

(a) Status of match requests.

(b) Maximum weight match (MWM).

(c) Maximal weight match.

FIGURE 6.5
Example of maximum and maximal weight matching.

MWM schemes have been proved to achieve 100% throughput under admissible traffic [110]. This maximum throughput was proved by showing stability in the queue occupancy of VOQs. Stability of a queue here means that the queue occupancy does not grow indefinitely. Therefore, it is appealing to consider matching schemes that may mimic the matching results of MWM schemes with, however, lower complexity. In the following sections, we present some existing maximal-weight matching schemes.

6.6.1   Iterative Longest Queue First (iLQF)

The iterative longest queue first (iLQF) matching scheme performs maximal-weight matching. iLQF uses queue occupancy of VOQs as weights for port selection. The scheme is executed as a parallel matching process, using three phases: request, grant, and accept. The operation of iLQF is described as follows:

(1) An input sends requests to all outputs for which it has a packet.

(2) Each output arbiter selects the request with the largest weight among all received. Ties are broken arbitrarily (the port with the smallest index number maybe selected, as an example).

(3) Each input arbiter selects the grant with the largest weight among all received. Ties are broken arbitrarily.

The three steps are repeated among unmatched ports in each iteration. The scheme may be performed up to N iterations. Figure 6.6 shows an example of the iLQF scheme. Figure 6.6(a) shows the status of the queues, where the numbers are the weights in each request (number of cells in the corresponding VOQ). Figure 6.6(b) shows the grant phase, where the output arbiters select the largest weighted request and issue a grant towards the input arbiters. Figure 6.6(c) shows the accept phase and the resulting match.

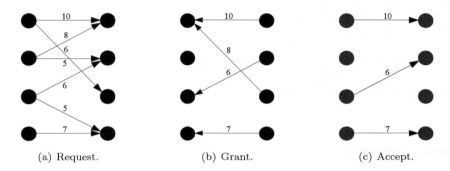

(a) Request.   (b) Grant.   (c) Accept.

FIGURE 6.6
Example of maximal weight matching.

iLQF serves the longest queue [110]. In general, to achieve the largest weight match, iLQF may require to perform up to N iterations. If the initial occupancy of a VOQ is the largest among all in the switch, and this queue continues to experience cell arrivals, iLQF will continuously serve such queue. This leads to starvation of other queues. This starvation may last as long as the condition remains. This means that some cells will experience an indefinite queueing delay and fairness may not be achieved.

6.6.2 Iterative Oldest Cell First (iOCF)

The iterative oldest cell first (iOCF) scheme operates in a similar way as the iLQF scheme does. Instead of considering the queue occupancy of a VOQ as the matching weight, iOCF uses the time a cell (or packet) has been waiting in the VOQ. This scheme has the advantage of ridding the scheduler from queue starvation. In iLQF, a queue may monopolize the scheduler service as long as the queue keep being the longest queue. This scenario may occur if the queue is the longest and keeps receiving traffic. In contrast, iOCF considers the cell that has waited the longest time in the queue as one with the highest priority.

This scheme also faces some challenges. Differently from iLQF, keeping the time a cell waits in a queue is more complex as any unit time passing has to be accounted for. This means that a large time counter needs to be used before the time count wraps around (i.e., starts from zero again). A wrapped-around value may confuse the queue selection process. A large number of bits are needed by the counter to track the time. Under uniform traffic, this scheme may not perform better than iLQF.

The description of the matching process of iOCF is similar to that of iLQF, where the waiting time is used instead of the queue occupancy. A feature shared by weighted-based matching schemes is that arbiters don't need to remember the last match. This feature may simplify the implementation complexity of the scheme, which uses several comparators to decide the most preferred port in a weighted-selection arbiter. Also, the number of iterations that this scheme performs depends on the allowed time budget, as in any other iterative scheme. The largest number of iterations this scheme may perform is N.

This scheme can achieve high throughput under nonuniform traffic in the maximal-weight matching version. Its MWM version has been adopted for achieving 100% under admissible traffic [113] as this is a no-starvation policy.

6.7 Size-Matching Schemes

Size-matching schemes are aimed at achieving the largest number of matches, independently of the queue occupancy status or age of cells at VOQs. Some of the objectives of this group of matching schemes are:

- Transfer the largest number of cells from inputs to outputs.

- Achieve the highest throughput, if not 100%, under uniform traffic.

- Achieve the two objectives above with the smallest number of iterations, if not a single one.

Size-matching schemes include parallel iterative matching (PIM) and round-robin based schemes, such as iterative round-robin matching (iRRM), and iterative round-robin with Slip (iSLIP). PIM selects ports randomly; it does not have a priority list, nor requires remembering which port was last matched. Round-robin-based schemes aim to provide a fair service to inputs and outputs. To do this, they follow a fixed schedule and a pointer to remember the last match.

6.7.1 Parallel Interactive Matching (PIM)

PIM is a size-based matching scheme and uses no pointers nor requires remembering the past matches [8]. In theory, it is a very simple scheme and shows how parallel matching can be applied for fast contention resolution. This simplicity, however, impacts the achievable performance; it provides limited throughput under uniform Bernoulli uniform traffic. This scheme follows three steps (initially, all inputs and outputs are unmatched):

Step 1. Each unmatched input sends a request to every output for which it has a queued cell.

Step 2. If an unmatched output receives any requests, it randomly chooses one to grant. The output notifies each input whether its request is granted.

Step 3. If an input receives any grants, it randomly chooses one to accept and notifies that output.

PIM can achieve high throughput under uniform traffic in N iterations, but the achieved throughput under fewer iterations is low.

6.7.2 Iterative Round-Robin Matching (iRRM)

Round-robin arbitration provides fair service among contending ports (or queues). The iRRM is an iterative matching scheme using round-robin as selection policy. In this scheme, inputs (and outputs) follow a sorted order. Each input and output has an arbiter, called input and output arbiter, respectively. To remember the last granted port, an arbiter has a pointer. The pointer of an input arbiter is denoted as a_i and the pointer of an output arbiter is denoted as g_j. These pointers mark the port with the highest matching priority. iRRM is performed as described by the following three steps:

Request. Each input sends a request to every output for which it has a queued cell.

Grant. If an output receives any requests, it chooses the one that appears next in a fixed, round-robin schedule starting from the port indicated by the pointer. The output arbiter g_j notifies each input whether or not its request was granted. The output pointer is incremented (modulo N) to one location beyond the granted input.

Accept. If an input receives a grant, it accepts the one that appears next in a fixed round-robin schedule, starting from the pointer position. The input pointer to the highest priority element of the round-robin schedule is incremented (modulo N) to one location beyond the accepted output.

Figure 6.7 shows an example of iRRM. The occupancy of a VOQ is the number of packets at input i for output j and it is represented as $L(i, j)$. Figure 6.7(a) shows the initial state of VOQ occupancy and pointer position of input and output arbiters. During the request phase, Inputs 1, 2, and 4 send requests to the outputs they have cells for. All outputs receive one or more requests. In the grant phase (Figure 6.7(b)), outputs (arbiters) select the request closer to the pointer position and next in the round-robin order. For example, Output 3 receives three requests, but the pointer for this arbiter indicates that Input 1 is the most preferred input port, so it grants the request from Input 1. Figure 6.7(c) shows the accept phase, where only two inputs are matched. Note that after each selection phase, pointers update their position; output pointers are updated in the grant phase, and input pointers are updated in the accept phase.

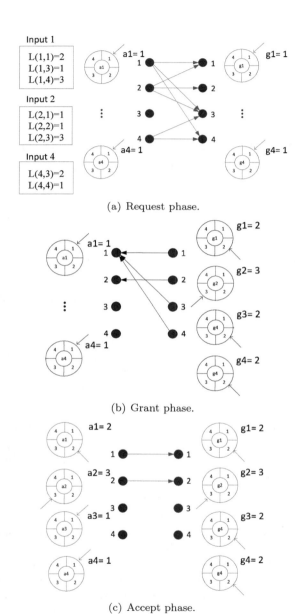

(a) Request phase.

(b) Grant phase.

(c) Accept phase.

FIGURE 6.7
Example of *i*RRM.

6.7.3 Round-Robin with Slip (*i*SLIP)

Despite being based on round-robin selection, *i*RRM is not able to achieve 100% throughput under uniform traffic using a single iteration, as arbiters keep pointing to the same ports, whether they get matched or not. This phenomenon is called pointer synchronization. To overcome this drawback of *i*RRM, the iterative round-robin with slip, or *i*SLIP, was proposed. To desynchronize the pointers, *i*SLIP updates them only if an issued grant is accepted in the accept phase of the matching process.

The operation of *i*SLIP is described as follows:

Request. Each input sends a request to every output for which it has a queued cell.

Grant. If an output receives any requests, it chooses the one that appears next in a fixed round-robin schedule starting from the port indicated by the pointer. The output arbiter g_j notifies each input whether or not its request was granted.

Accept. If an input receives a grant, it accepts the one that appears next in a fixed, round-robin schedule starting from the highest priority element. The input pointer is incremented (modulo N) to one location beyond the accepted output. The output pointer to the highest priority element of the round-robin schedule is incremented (modulo N) to one location beyond the granted input for accepted grants.

*i*SLIP may be run in multiple iterations. In that case, pointers are updated in the first iteration, only. Unmatched ports are matched in the following iterations. Figure 6.8 shows an example of *i*SLIP, running one iteration. This example is similar to the one for *i*RRM. So the request and grant phases (Figures 6.8(a) and 6.8(b)) are similar to those in *i*RRM, except that the output pointers don't update their position yet. In the accept phase (Figure 6.8(c)), input arbiters select their grant and issue the accepts for the granted outputs. In this phase, input and output pointers of the matched ports update their positions.

As expected, *i*SLIP not only achieves 100% throughput under uniform traffic but also achieves this by a single iteration. Adopting this single iteration in a matching scheme saves time and allows to achieve higher data rates than those schemes that require multiple iterations to achieve equivalent performance.

6.7.4 Dual Round-Robin Matching (DRRM)

The order in which the three-step matching process schemes like *i*SLIP and iRRM use may be reversed (i.e., where inputs make the first selection) but also minimized. If the inputs send a single request, then the accept phase

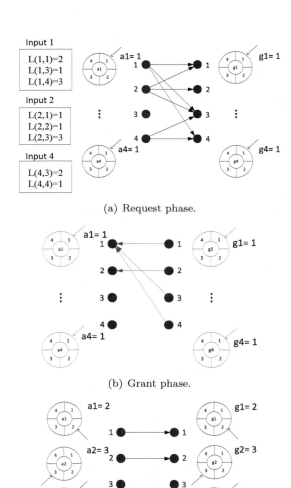

(a) Request phase.

(b) Grant phase.

(c) Accept phase.

FIGURE 6.8
Example of iSLIP (1 iteration, or 1SLIP).

of the three-step matching process can be removed. DRRM uses these two observations to simplify the matching complexity [30].

DRRM is used in packet switches, where each input has N VOQs. An input arbiter, r_i, at each input selects a nonempty VOQ according to the round-robin order. After the selection, each input port sends one request, if any, to an output arbiter, r_j. An output arbiter at each output receives up to N requests and chooses one of them based on the round-robin policy, and sends a grant to the chosen input port. The DRRM scheme is described as follows:

Step 1: Request. Each input sends an output request corresponding to the first nonempty VOQ in a fixed round-robin order, starting from the current position of the pointer. The pointer remains at that nonempty VOQ if the selected output is not granted in Step 2. The pointer of the input arbiter is incremented by one location beyond the selected output if, and only if, the request is granted in Step 2.

Step 2: Grant. If an output receives one or more requests, it chooses the one that appears next in a fixed round-robin schedule starting from the current position of the pointer. The output notifies each requesting input whether or not its request is granted. The pointer of the output arbiter is incremented to one location beyond the granted input. If there are no requests, the pointer remains where it is.

Figure 6.9 shows an example of DRRM. In this example, Input 1 has cells destined for Outputs 1 and 2, Input 2 has cells for Outputs 1, 3, and 4. Input 3 has cells for Output 3, and Input 4 has cells for Outputs 2 and 3. Because $r_1 = 1$, Input 1 issues a request to Output 1. In a similar way, Inputs 2, 3, and 4 issue requests to Outputs 3, 3, and 2, respectively. In the grant phase, each output selects at most one input. Output 1 grants the request from Input 1 and the pointer remains pointing to Input 2, as this is one position beyond the granted input. Output 2 grants the only request, from Input 4, and the pointer moves to Input 1. Output 3 grants Input 3 as $g_3 = 3$ and the pointer is moved towards Input 4.

DRRM, as *i*SLIP, achieves 100% throughput under uniform traffic in a single iteration. Pointer desynchronization is also achieved with this scheme. At the same time, the simple round-robin selection policy adopted inhibits DRRM from achieving high switching performance under nonuniform traffic patterns. DRRM may also adopt multiple iterations [129].

6.7.5 Round-Robin with Exhaustive Service

While schemes that perform matching every time slot show that the selection policy may perform differently under different traffic patterns, this approach also demands a high rate in decision making (e.g., every time slot). This change of matching seems to make one think that the ability for a scheme to make

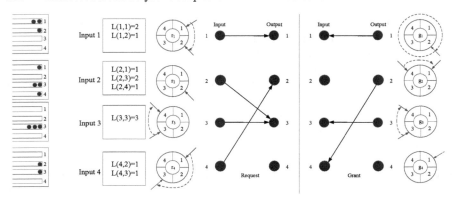

FIGURE 6.9
Example of DRRM.

complex and rapid matchings may help to increase the switch throughput. However, other schemes show that such rate of matching (and for reconfiguration of the switch fabric) may not be necessary.

The round-robing scheme with exhaustive service, or EDRRM, shows that it may be convenient to keep *successful* matchings in place [105]. A successful matching is that where an input is matched to an output, and a cell (or packet) is forwarded through that connection (a counter example is one where a matching between an input and output may be led by other port matchings in the switch, and yet no cell is forwarded through it as the number of cells for that connection is very small or zero). The EDRRM algorithm is described as follows:

Step 1: Request. Each input moves its pointer to the first nonempty VOQ in a fixed round-robin order, starting from the current position of the pointer, and sends a request to the output corresponding to the VOQ. The pointer of the input arbiter is incremented by one location beyond the selected output if the request is not granted in Step 2, or if the request is granted and after one cell is served this VOQ becomes empty. Otherwise, the pointer remains at that (nonempty) VOQ.

Step 2: Grant. If an output receives one or more requests, it chooses the one that appears next in a fixed round-robin schedule starting from the current position of the pointer. The output notifies each requesting input whether or not its request is granted. The pointer of the output arbiter remains at the granted input. If there are no requests, the pointer remains where it is.

Figure 6.10 shows an example of EDRRM. Arbiter for Input i is r_i and arbiter for Output j is g_j. The arbiters have a pointer pointing towards the highest priority port (i.e., the most preferred port for matching). Here, Input 1 has cells for Outputs 1 and 2, Input 2 has cells for Outputs 1, 3, and 4, Input

3 has cells for Output 3, and Input 4 has cells for Outputs 2 and 3. Initially, r_1 points to Output 1, r_2 and r_3 point to Output 3, and r_4 points to Output 2. Similarly, g_1, g_2, g_3, and g_4 point to Input 2, 4, 3, and 1, respectively. As Input 1 has multiple cells for Output 1, r_1 selects Output 1 for sending a request. In a similar way, r_2 and r_3 select Output 3, and r_4 selects Output 2, and requests are sent. After receiving the requests and in consideration of the pointer positions, g_1, g_2, and g_3 select Input 1, Input 4, and Input 3, respectively. Pointers r_1 and r_3 remain at their position as they were granted and have more cells for that output (no new match is needed). Here, r_2 updates its pointer position to Output 4 as the request was denied. The pointer position of r_4 is also updated as the queue for Output 2 at Input 4 becomes empty.

Because g_1 grants Input 1, the pointer position moves from Output 2 to Output 1, the pointer of g_2 remains at Input 4 (keeps the match), the pointer of g_3 remains pointing at Input 3, and g_4 remains the same (as it didn't receive any matching request).

EDRRM can provide high throughput under nonuniform traffic patterns but it is not able to keep the high switching performance under uniform traffic as did previous round-robin-based matching schemes. Furthermore, some starvation issues are also of concern in EDRRM.

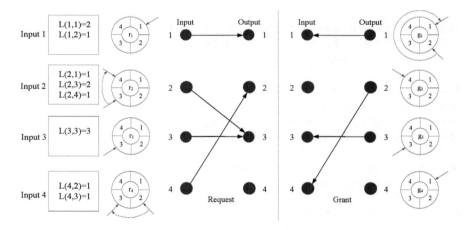

FIGURE 6.10
Example of EDRRM.

6.8   Frame-Based Matching

Sending multiple cells after every match may help to achieve high throughput under traffic with nonuniform distributions and through the use of size matching. The train of cells may be seen as a frame. Switching cells in a frame may reduce the frequency in which the fabric is reconfigured.

6.8.1   Frame Size Occupancy-Based Schemes

The exhaustive service model presented in EDRRM showed that servicing a queue for too long can present starvation issues, and therefore, the scheme is unable to achieve 100% throughput under uniform traffic. To address that, framed-based schemes where the frame size is set to the number of cells in a VOQ at a given time were proposed [146, 147]. These schemes are the unlimited frame-size occupancy-based PIM (uFPIM) and unlimited framed-size occupancy-based round-robin matching (uFORM) schemes.

We use the following definitions in the description of the uFPIM and uFORM matching schemes.

Frame. A frame is associated to a VOQ. A frame is the set of one or more cells in a VOQ that are eligible for dispatching. Only the HoL cell of the VOQ is eligible per time slot.

Captured frame size. At the time tc of matching the last cell of the frame associated to $VOQ(i, j)$, the next frame is assigned a size equal to the occupancy of $VOQ(i, j)$. Cells arriving in $VOQ(i, j)$ at time td, where $td > tc$, are not considered for matching until the current frame is totally served and a new frame is captured. We call this captured frame as it is the equivalent of having a snapshot of the VOQ occupancy at time tc, where the occupancy determines the frame size.

On-service status. A VOQ is said to be in on-service status if the VOQ has a frame size of two or more cells and the first cell of the frame has been matched. An input is said to be on-service status if the status of a VOQ becomes on.

Off-service status. A VOQ is said to be in off-service status if the last cell of the VOQ's frame or no cell of the frame has been matched. Note that for the frame size of one cell, the corresponding VOQ is off-service during the matching of its only cell.

6.8.1.1   Unlimited Frame-Size Occupancy based PIM (uFPIM)

In the uFPIM scheme, each VOQ is associated with a capture-frame counter, denoted as $(CF_{i,j})$, and an on/off-service status flag, denoted as $(F_{i,j})$. uFPIM follows three steps as in the PIM scheme:

- **Request.** Nonempty on-service VOQs send a request to their destined output. Nonempty off-service VOQs send a request to their destined outputs if input i is off-service.

- **Grant.** If an output arbiter g_j receives any requests, it chooses a request from the on-service VOQ (also called an on-service request) in a random fashion. If there is no on-service request, the output arbiter chooses an off-service request in a random fashion.

- **Accept.** If the input arbiter a_i receives any grants, it accepts one on-service grant in a random fashion. If there is no on-service grant, the arbiter chooses an off-service grant in a random fashion. The CF counter updates the value according to the following: If the input arbiter a_i accepts a grant from a_j, and if:

  (i) $CF_{i,j}(t) > 1$: $CF_{i,j}(t+1) = CF_{i,j}(t) - 1$ and this VOQ is set as on-service, $F_{i,j} = 1$.

  (ii) If $CF_{i,j}(t) = 1$: $CF_{i,j}(t+1)$ is assigned the occupancy of $VOQ(i,j)$, and $VOQ(i,j)$ is set as off-service, $F_{i,j} = 0$.

Figure 6.11 shows an example of matching with uFPIM. The CF values also show the VOQs' content, the captured-frame sizes, and the service status of each VOQ. In the request phase, Inputs 1, 2, and 3 send off-service requests to all outputs they have at least a cell for. Input 4 sends a single on-service request to Output 1, as the off-service VOQ is inhibited as described in the scheme. The output and input arbiters select a request by service status and in a random fashion among all requests of the same service status, as shown by the grant and accept steps. Output 1 selects the on-service request from Input 4 over the off-service request from Input 2. After the match is performed, the CF values are updated to $CF_{2,3}(t+1) = 1$, $CF_{3,2}(t+1) = 1$, and $CF_{4,1}(t+1) = 4$. Also, at time slot $t+1$, the status of three VOQs becomes on-service.

The observed throughput of uFPIM is 100% under uniform traffic and about 99% under unbalanced traffic [139], diagonal, and power-of-two (Po2) traffic patterns. Here, the frame size is defined by the traffic pattern received and the rate in which cells arrive to a VOQ. The capture frame concept limits the frame size to the existing cells in the queue at the time the frame is captured (this is analogous to taking a photograph of the status of the VOQ at that time). The captured frame concept also improves the fairness of the scheme as no new arriving cells may be served before some other older cells.

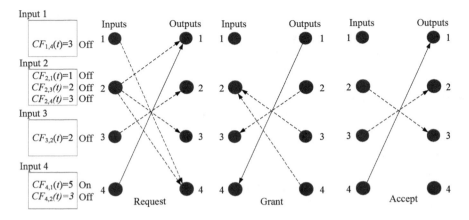

FIGURE 6.11
Example of uFPIM.

6.8.1.2 Unlimited Framed-Size Occupancy based Round Robin Matching (uFORM)

A round-robin version of a frame-size occupancy based scheme is represented by the uFORM scheme. The throughput measured of uFORM is 100% under uniform traffic and about 99% under unbalanced, diagonal, and PO2 traffic patterns. It also achieves 100% throughput under Chang's and asymmetric traffic patterns. uFORM follows request-grant-accept steps as in uFPIM, and uses round-robin selection instead of random-based selection. The matching process is as follows:

Step 1: Request. Nonempty on-service VOQs send a request to their destined outputs. Nonempty off-service VOQs send a request to their destined outputs if input i is off-service.

Step 2: Grant. If an output arbiter a_j receives any requests, it chooses a request from the on-service VOQ (also called an on-service request) that appears next in a round-robin schedule, starting from the pointer position. If there is no on-service request, the output arbiter chooses an off-service request that appears next in a round-robin schedule, starting from its pointer position.

Step 3: Accept. If the input arbiter a_i receives any grants, it accepts an on-service grant in a round-robin schedule, starting from the pointer position. If there is no on-service grant, the arbiter chooses an off-service grant that appears next in a round-robin schedule starting from its pointer position. The input and output pointers are updated to one position beyond the matched one. In addition to the pointer update, the CF counter updates the value according to the following: if the input arbiter a_i accepts a grant from a_j, and if:

i) $CF_{i,j}(t) > 1 : CFi,j(t+1) = CF_{i,j}(t) - 1$ and this VOQ is set as on-service, $F_{i,j} = 1$.

ii) If $CF_{i,j}(t) = 1$: $CF_{i,j}(t+1)$ is assigned the occupancy of $VOQ(i, j)$, and $VOQ(i, j)$ is set as off-service, $F_{i,j} = 0$.

Figure 6.12 shows an example of uFORM in a 4×4 switch. In this example, the contents of the VOQs are the same as that of the uFPIM example (Figure 6.12). The pointers of the input and output arbiters are positioned as shown in the request phase. The off-service inputs send a request to all outputs they have a cell for. In the grant phase, the output arbiters select the request according to the request status and the pointer position. Output 1 selects the on-service request over the off-service request. Output 4 receives two off-service requests, and selects Input 2 because that input has higher priority according to the pointer position. Outputs 2 and 3 receive a single off-service request. Therefore, the requests are granted. In the accept phase, Input 2 selects Output 3 according to the pointer position. Input 3 accepts the single grant issued by Output 2. Input 4 accepts the single grant, issued by Output 1. In this case, the match of uFORM is the same as in the uFPIM example. Therefore, the CF values and service status are updated as in that example. Note that the input and output arbiters for the on-service ports (Input 4 and Output 1) are updated, but since the service status takes higher precedence, the pointer position in this case becomes secondary in the selection process.

uFORM, as uFPIM, shows that the capture frame concept improves the switching performance of size matching even under nonuniform traffic patterns. uFORM achieves slightly higher performance than uFPIM because the desynchronization of arbiters achieved in uFORM allows to forward a large number of cells. The *captured frame* concept is particularly useful for building large switches, such as IQ Clos-network switches [106].

Schemes that avoid rapid or time-slot-based reconfiguration have gained recent interest as they relax the timing in which a switch operates [68, 168, 173].

6.9 Output-Queued Switch Emulation

Achieving the maximum throughput of 100% may not be enough to guarantee high switching performance. An OQ switch achieves such throughput and, in turn, the best delay performance. That is, a cell in an OQ switch may spend the smallest time in a queue. In an OQ switch, cells are stored at the output queues after they traverse the switch fabric. This means that cells are available at the output after arrival and this lets us select which cells are dispatched from the switch.

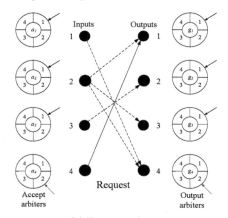

(a) Request phase.

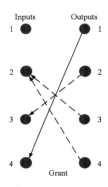

(b) Grant phase.

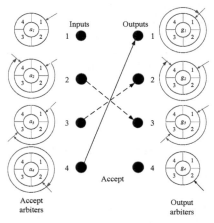

(c) Accept phase.

FIGURE 6.12
Example of uFORM.

With this motivation, it has been of interest in finding the conditions for which an IQ switch would be able to emulate the operation of an OQ switch and, therefore, achieve similar performance. The scheme used to select packets from the inputs and the used speedup are some of the conditions. The use of speedup shapes the switch as a combined input-output queued (CIOQ) switch. In a CIOQ switch with speedup S, cells are stored in an output queue with a speed S times faster than the line rate. The memory used to implement such a queue is required to work at that speed.

Figure 6.13 shows an example of how speedup can be considered in a CIOQ switch. A switch with no speedup may accommodate a matching phase (selection of the configuration of the switch) and the transmission of the matched cells in a single time slot (although in a more realistic implementation the matching and transmission phases are pipelined, here we show it in this way for the sake of clarity). Under a speedup, the number of matching and transmission phases occurs S times in a time slot. The bottom part of this figure shows an example for $S = 2$, and in that case the maximum number of cells transmitted is $2N$. The potential of speedup to transmit a larger number of cells empowers the scheme to outperform a version of it without speedup. However, memory is lagging current data rates, and speedup exacerbates the challenge for building such switches.

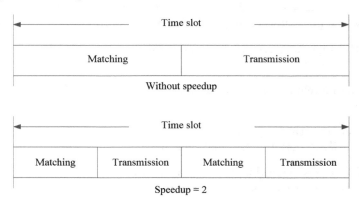

FIGURE 6.13
Example of timing for speedup of two. There are two matching phases and two transmissions.

A few algorithms have been proposed and analyzed to show that an IQ switch may emulate an OQ switch. The most urgent cell first algorithm (MUCFA) [134] follows a mirror OQ switch used to find out the order and time slots in which incoming cells would leave the switch. MUCFA in an IQ switch may use a speedup of four to achieve this. The lowest occupancy output first algorithm (LOOFA) [95] was also proposed and showed that a speedup of two would be sufficient to achieve such performance. In another study, the last in highest priority (LIHP) [38] scheme was conjectured that it may take a speedup of $2-\frac{1}{N}$ to emulate an OQ switch.

6.10   Exercises

1. Show an example of HoL blocking for Inputs 2 and 4 and Output 2 for a 4×4 switch.

2. Show an example of a set of requests (i.e., occupancy in VOQs) in a 4x4 switch such that the MWM match is equivalent to the maximal-weight match.

3. Show an example of a set of requests (i.e., occupancy in VOQs) in a 4x4 switch such that the MWM match is different from the maximal-weight match.

4. Consider an IQ 3x3 switch with VOQs. The switch has the following occupancies: L(1,1)=2, L(1,2)=2, L(1,3)=1, L(2,1)=2, L(2,2)=2, L(2,3)=1, L(3,1)=2, L(3,2)=1 and the pointers are all initially pointing to indexes 1. Show the matching results in the following two time slots using 1RRM.

5. Show the matching results in Exercise 4 but using 1SLIP as matching scheme.

6. Show the matching results in Exercise 4 but using EDRRM as matching scheme.

7. Investigate how the throughput of a FIFO IQ switch was analyzed and found to be 58.6%. Report the information source and a brief description.

8. Solve Exercise 4 again but using two iterations every time slot. Show the packets that are switches to the outputs.

9. Show the matching of Exercise 4 but using uFORM for the following four time slots.

10. Show the packets leaving a switch with the occupancies as in Exercise 4, where the switch uses a speedup of two, for two time slots, using iRRM.

11. Argue what the possible implications are on the operation and performance of a switch that would occur if a time slot duration is extended to accommodate a slow scheduler.

7

Shared-Memory Packet Switches

CONTENTS

Switches with shared memory are attractive as memory may be better utilized by all the switch ports. Switching of packets from inputs to outputs in a shared-memory switch can follow the behavior of an output-buffered switch and even offer higher memory efficiency. In a shared-memory switch, packets coming from different inputs and going to different outputs are all stored in the same memory. Therefore, the design of these switches must take into account the required memory speed and policies to make sharing of memory practical and fair. Memory speed is a major challenge for building a shared-memory switch. In this chapter, we discuss architectures of shared-memory switches and policies for managing the sharing of memory.

7.1 Introduction

Shared-memory switches were first proposed to resolve the output contention for practical implementation of Asynchronous Transfer Mode (ATM) [115] switches [59, 98]. In these switches, memory is used as part of the switch fabric and to store packets that may lose output contention. In a shared-memory switch with a single memory block, the memory is shared by both input and output ports. Therefore, to keep up with the link rates, the inputs and output ports should be allowed to receive and forward packets at the link rates. This means, for an $N \times N$ switch, the shared memory may be accessed up to N^2 times during the time a packet is transmitted under multicast traffic. In other words, the internal parts of the switch, including the memory, are required to run N^2 faster than the transmission speed of the input or output ports, in the worst-case scenario. This speed difference between the links and the internals of the switch is generally called memory speedup. It is then easy to see that speedup grows quickly as N grows for a shared-memory switch. Consequently, this memory speedup becomes a performance bottleneck of a shared-memory switch with a large number of ports because memory is a speed-lagging component.

The following sections overview different shared-memory switches and schemes that mitigate the memory speedup problem. Important buffer management schemes to efficiently control the memory access while maintaining high throughput and low latency of the switch are introduced.

In the switches described in this chapter, we continue to follow the mainstream practice of segmenting incoming variable-size packets into fixed-length packets, called cells, at the ingress side of a switch. Packets are reassembled at the egress side before they depart from the switch. Each cell takes a fixed amount of time, or time slot, to transmit from an input to an output. The terms cells and time slots are used interchangeably for indication of time.

7.2   Basics of a Shared-Memory Packet Switch

The basic shared-memory packet switch has a block of memory shared by input and output ports as shown in Figure 7.1. With the growing popularity of multimedia applications, it is desirable that a packet switch is capable of supporting unicast and multicast traffic. Memory access under these traffic types may be different and so are their speedup requirements.

Under unicast traffic, an $N \times N$ shared-memory switch may need to perform N writes, one write into memory per input port, and N reads, one per output port, in the memory block. These numbers of access are performed if a packet is written/read in a single memory access (using L-bit wide memory bus where L is the cell size). In this case, memory requires a speedup of $2N$. Note that if the writing and reading operations in such a switch require multiple memory accesses, these must be accounted for and the result is a different speedup.

If multicast traffic is considered, the required memory speedup would depend on how packets are replicated and stored in the switch and, in turn, in the memory block. For example, let's consider that these multicast packets are replicated as unicast packets and, therefore, stored in unicast queues.

In addition, let's consider that memory writes and reads are not processed at the same time (they are not overlapped or processed at the same time). In this case, the switch must perform up to N^2 writes and up to N reads in one time slot under the worst-case scenario, where each input port has to replicate N packets. This results in a speedup of $N^2 + N$. Therefore, the memory speedup under multicast traffic grows faster than under unicast traffic as N grows in the example switch.

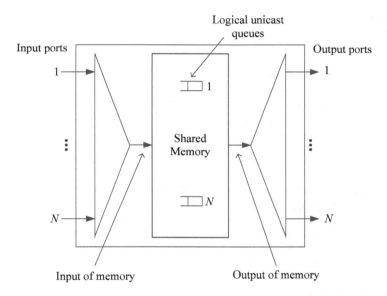

FIGURE 7.1

$N \times N$ shared-memory switch.

The speedup requirement is one of the reasons why shared-memory packet switches are not scalable. To reduce the required memory speedup, the number of accesses to the memory block has to be reduced. One possible strategy is to use multiple blocks of memory, where each block may be assigned to a small set of ports. Figure 7.2 shows an example of a switch with two memory blocks, each block dedicated to $N/2$ output ports and shared by all input ports. The drawback of dedicating memory to a particular number of ports is that the level in which the memory is shared is small (it is only shared by those ports to which memory is dedicated), but the complexity of access management of the shared memory also decreases. As the cost of memory density decreases, it may be more effective to partition the blocks of memory in a shared-memory switch.

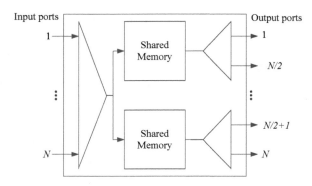

FIGURE 7.2
$N \times N$ shared-memory switch with two memory blocks.

In the two-block shared memory switch, the speedup for unicast traffic is $\frac{3}{2}N$, as the switch may need to perform up to N writes and $N/2$ reads in a time slot. As for multicast traffic (with the switch features as described above), the required memory speedup is $N^2 + N/2$. Therefore, there is a small reduction on the required speedup. This reduction is more noticeable in the case of unicast traffic.

7.2.1   Shared-Memory Packet Switches with Multiple Memory Blocks

The access to the shared-memory block in a shared-memory packet switch must be managed to allow a dynamic allocation of memory to input/output ports so as to efficiently use that memory. This utilization of memory requires logical queues allocated to each input/output port. It is then expected that a switch must include a module for memory management. Figure 7.3 shows a management block in a shared-memory switch.

Using a number of memory blocks, each shared by a subset of input/output ports, may reduce the speed in which the memory needs to operate and the complexity of the memory allocation scheme. However, reducing the level of sharing may require a larger total amount of memory.

A shared-memory switch that has a single memory block uses a possible largest speedup. In contrast, a switch using a crossbar switch fabric with buffered crosspoints may use no speedup. Both of these two switches can be considered as switches with internal memory. A buffered crossbar switch may then use shared memory in the crosspoint buffers to increase the efficiency of memory usage.

To reduce the amount of memory in a buffered crossbar using crosspoint buffers, the shared-memory crosspoint buffered (SMCB) switch was proposed [51, 52, 53, 141, 142]. In an SMCB switch, a crosspoint buffer is shared by m

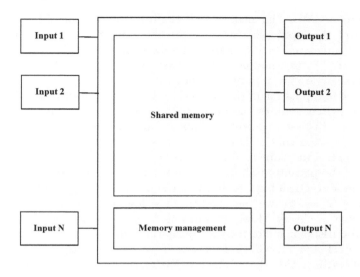

FIGURE 7.3
$N \times N$ shared-memory switch with memory management block.

inputs, where $2 \leq m \leq N$. This section presents three SMCB switches and discusses their operation and complexity.

7.3 Shared-Memory Crosspoint Buffered Switch with Memory Allocation and Speedup of m (SMCB×m)

The shared-memory crosspoint buffered switch with memory allocation and speedup of m, SMCB×m, has N VOQs at each input, and a crossbar with N^2 crosspoints and $\frac{N^2}{m}$ crosspoint buffers [51]. A crosspoint in the buffered crossbar that connects input i to output j is denoted as $CP(i,j)$, as in a combined input-crosspoint buffered (CICB) switch. The buffer, or shared-memory buffer, SMB, is shared by m crosspoints, therefore $2 \leq m \leq N$. When $m = 2$, the buffer shared by $CP(i,j)$ and $CP(i',j)$, where $1 \leq i' \leq N$ and $i \neq i'$, that stores cells for output j is denoted as $SMB(q,j)$, where $1 \leq q \leq \lfloor \frac{N}{2} \rfloor$. The size of $SMB(q,j)$, k_s, is given in the number of cells it can store.

Each VOQ has a service counter to count the number of cells forwarded to the buffered crossbar, and a counter limit, $C_{i,j}^{max}$, to indicate the maximum number of cells that $VOQ(i,j)$ can forward to the corresponding SMB. A sharing control unit (SCU) at each SMB determines the size of the partition assigned to an SMB. The memory partitioning is in function of the occupancy of VOQs of the inputs sharing the corresponding SMB. The occupancy of $VOQ(i,j)$ is denoted as $Z_{i,j}$. Table 7.1 shows the buffer allocation, indicated by counter $C_{i,j}^{max}$, in function of $Z_{i,j}$ of each VOQ. The units of the allocation and occupancy are given in number of cells.

Because m inputs might demand access to the shared memory at the same time, this switch requires the memory to work with a speedup of m, where $m \geq 1$. A credit-based control mechanism, in combination with the dynamic memory allocation that SCU manages, is used to avoid buffer overflow. With the flow control mechanism, an input is kept from sending more cells to the SMB than the permitted allocation. To minimize the speedup of the shared memory in a practical implementation, the number of inputs sharing a crosspoint buffer is set to two (i.e., $m=2$). This description considers an even N for the sake of clarity. However, an odd N can also be adopted (with one dedicated crosspoint buffer for the nonsharing input).

Figure 7.4 shows the SMCBxm switch with two inputs sharing the crosspoint buffers (i.e., SMCBx2). The SMCBx2 switch uses random selection for input and output arbitrations. The switch works as follows: When a cell arrives in $VOQ(i,j)$, a request is sent to the corresponding SCU. Based on $Z_{i,j}$ of the inputs sharing the SMB, the SCU at every SMB sets up the size of the memory partition available for each input. The allocated amount of memory sets C^{max} and the flow control keeps track of the occupancy of the SMBs and notifies the inputs. A grant is sent from the SCU to the input to enable the forwarding of a cell, and the input dispatches a cell in the next time slot. The count of the service counter is increased by one when a cell from $VOQ(i,j)$ is sent to $SMB(q,j)$ and reduced by one when a cell of $VOQ(i,j)$ at $SMB(q,j)$ is dispatched to the output. When the count of the service counter reaches $C_{i,j}^{max}$, $VOQ(i,j)$ is inhibited from sending more cells to the SMB. After cells are stored at the SMBs, the output arbiter selects a cell to be forwarded to the output. The selected cell is sent to the output in the next time slot.

Figure 7.5 shows an example of a 4×4 SMCBx2 switch with $k_s = 2$ cells. Inputs 1 and 2 share $SMB(1,j), 1 \leq j \leq 4$. Inputs 3 and 4 share $SMB(2,j), 1 \leq j \leq 4$. At time slot T, $VOQ(1,1)$ holds Cells A and B, $VOQ(2,1)$ holds Cells C and D, $VOQ(3,1)$ holds Cells E and F, and $VOQ(4,4)$ holds and Cells G and H, as Figure 7.5(a) shows. Because the occupancy of $VOQ(1,1)$ and $VOQ(2,1)$ is the same, $SMB(1,1)$ allocates $k_s/2 = 1$ cell to each VOQ, and $C_{1,1}^{max} = C_{2,1}^{max} = 1$. Because $VOQ(3,4)$ and $VOQ(4,1)$ have no cells, $SMB(2,1)$ and $SMB(2,4)$ allocate their full capacity to $VOQ(3,1)$ and $VOQ(4,4)$, respectively, and $C_{2,1}^{max} = C_{2,4}^{max} = k_s = 2$ at this time slot.

At time slot $T + 1$, Cells A (from Input 0) and C (from Input 1) are forwarded to $SMB(1,1)$ and Cells E and G are forwarded to $SMB(2,1)$ and $SMB(2,4)$, respectively, as Figure 7.5(b) shows. The output arbiters at Outputs 1 and 4 select a cell to be forwarded to the output in a random fashion. At time slot $T + 2$, Cells A and G are forwarded to Outputs 1 and 4,

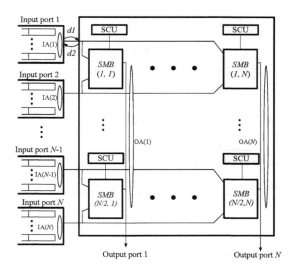

FIGURE 7.4
$N \times N$ SMCBx2 switch.

respectively, as shown in Figure 7.5(c). Flow control information is sent back to Inputs 1 and 4 to indicate the availability of the SMBs.

TABLE 7.1
Memory allocation of an SMB in the SMCBx2 switch.

$Z_{i,j}$	$Z_{i',j}$	$C_{i,j}^{max}$	$C_{i',j}^{max}$
0	0	$\lfloor k_s/2 \rfloor$	$\lfloor k_s/2 \rfloor$
$[0, RTT)$	$[0, k_s - RTT]$	$Z_{i,j}$	$k_s - Z_{i,j}$
$[RTT, \infty)$	0	k_s	0
$[0, RTT/2]$	$[0, RTT/2]$	$\lfloor k_s/2 \rfloor$	$\lfloor k_s/2 \rfloor$
$(RTT/2, \infty)$	$[0, RTT/2]$	$k_s - Z_{i',j}$	$Z_{i',j}$
$(RTT/2, \infty)$	$(RTT/2, \infty)$	$\lfloor k_s/2 \rfloor$	$\lfloor k_s/2 \rfloor$

7.4 Shared-Memory Crosspoint Buffered Switch with Input-Crosspoint Matching (mSMCB)

The shared-memory crosspoint buffered switch with input-crosspoint matching, mSMCB, uses a scheduling mechanism to control the access to memory by the inputs to avoid memory speedup. In this switch, only one input is allowed to access an SMB in a time slot. To schedule the access to the SMB among

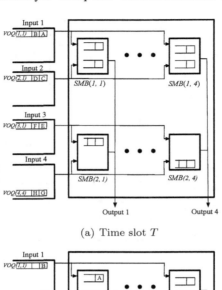

(a) Time slot T

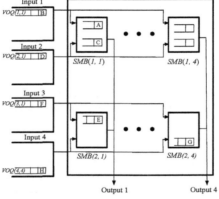

(b) Time slot $T + 1$

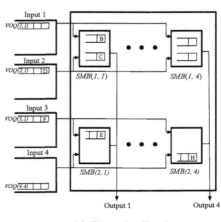

(c) Time slot $T + 2$

FIGURE 7.5
Example of a 4×4 SMCBx2 switch.

m inputs, an input-access scheduler, S_q, is used to match m nonempty inputs to up to N SMBs that may have room for storing at least one cell. Figure 7.6 shows the architecture of the mSMCB switch for $m = 2$, or 2SMCB switch. The size of an SMB, in number of cells that can be stored, is also denoted as k_s. There are $\frac{N}{m}$ S_qs in the buffered crossbar. A matched input is allowed to forward a cell to the corresponding SMB in the following time slot.

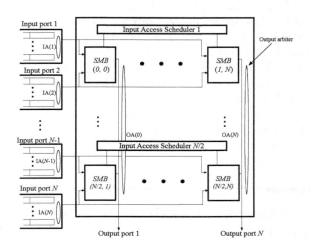

FIGURE 7.6
An $N \times N$ 2SMCB switch.

The matching in S_q follows a (3-phase) matching process as that used by some IQ switches [8, 111]. The matching scheme in S_q can adopt random selection, round-robin, or any other selection scheme. Here, we use random selection [8]. A credit-based flow control is used to monitor the available space in SMBs and avoid buffer underflow. Allocation of memory is not used in this switch as the matching process regulates the access to the SMBs because only an input is allowed to be matched to a single SMB, and an SMB is allowed to be matched to a single input.

At each output in the buffered crossbar, there is an output arbiter to select a cell from nonempty SMBs. An output arbiter considers up to two cells from each SMB, where each cell belongs to a different input. The output arbiter also uses random selection.

The mSMCB switch works as follows: Cells destined to output j arrive at $VOQ(i, j)$ and wait for forwarding. Input i notifies S_q about new cell arrivals. S_q selects the next cells to be forwarded to the crossbar by performing matching between inputs and SMBs. After a cell (or VOQ) is matched by S_q, the input is notified and sends the cell in the next time slot. A cell going from input i to output j enters the buffered crossbar and is stored in $SMB(q, j)$. Cells leave output j after being selected by the output arbiter.

Figure 7.7 shows an example of a 4×4 2SMCB switch with $k_s = 1$, where the occupancies of VOQs are the same as those in the previous example (Figure 7.5). It should be noted that this switch supports a smaller SMB than that in the SMCB$\times m$ switch. At time slot T, input access schedulers perform matching between inputs and SMBs. As shown in Figure 7.8, S_1 has two requests for $SMB(1,1)$ and grants Input 1 to access $SMB(1,1)$. S_2 has a request for $SMB(2,1)$ from Input 2 and one request for $SMB(2,4)$ from Input 3. Because both SMBs are available, both inputs are granted access to their corresponding SMBs. At time slot $T + 1$, Cell A is forwarded to $SMB(1,1)$, Cell E is forwarded to $SMB(2,1)$, and Cell G is forwarded to $SMB(2,4)$. The output arbiters at Outputs 1 and 4 select a cell, in a random fashion, to be forwarded to each corresponding output. The output arbiter at Output 1 selects Cell A and the output arbiter at Output 4 selects Cell G. Flow control information is sent back to Inputs 1 and 4 to indicate the availability of space at the corresponding SMBs. Cells A and G are forwarded to their destined outputs in time slot $T + 2$.

7.5   Shared-Memory Switch with Memory Shared by Output Ports

Shared-memory packet switches are a suitable alternative to support multicast services as content in the shared memory can readily be accessed by multiple ports without actual packet duplication. That is, these switches can potentially reduce communication overhead for multicasting.

In this section, we discuss an output-based shared-memory crosspoint-buffered packet switch, O-SMCB, that supports multicast traffic [53]. This switch requires less memory than a CICB switch to achieve comparable performance under multicast traffic. In addition, the O-SMCB does not require any speedup. Furthermore, the O-SMCB switch provides higher throughput under uniform and nonuniform multicast traffic models than an input-based SMCB switch, where two inputs share the crosspoint buffers when both switches use the same amount of memory.

For a fair comparison, the O-SMCB switch is provisioned with one multicast FIFO queue at each input. This switch has N^2 crosspoints and $\frac{N^2}{2}$ crosspoint buffers in the crossbar as each buffer is shared by two outputs. Figure 7.9 shows the O-SMCB switch. As in Section 7.4, a crosspoint in the buffered crossbar that connects input port i to output port j is denoted as $CP(i,j)$. The buffer shared by $CP(i,j)$ and $CP(i,j')$ that stores cells for output ports j or j', where $j \neq j'$, is denoted as $SMB(i,q)$, where $1 \leq q \leq \frac{N}{2}$. The following description considers an even N for the sake of clarity. However, an odd N can be used with one input port using dedicated buffers of 0.5 to 1.0 the size of an SMB. The size of an SMB, in number of cells that can be stored, is also k_s.

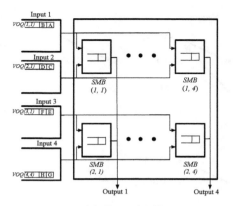

(a) Time slot T

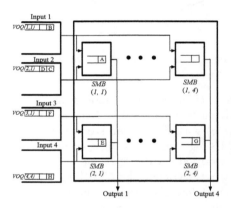

(b) Time slot $T + 1$

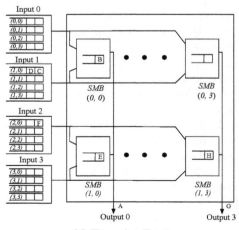

(c) Time slot $T + 2$

FIGURE 7.7
Example of a 4×4 2SMCB switch.

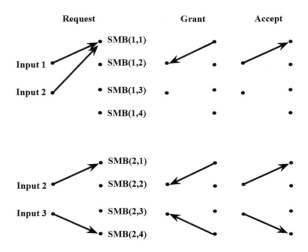

FIGURE 7.8
Matching at input access schedulers at time slot T.

Let us consider the case of minimum amount of memory at an SMB, when $k_s = 1$. This is equivalent to having 50% of the memory in the crossbar of a CICB switch. $SMB(i, q)$ with $k_s = 1$ can store a cell that can be directed to either j or j'. The SMB has two egress lines, one per output.

To avoid using speedup at SMBs, only one output is allowed to access an SMB at a time. The access to one of the $N/2$ SMBs by each output is decided by an output-access scheduler. The scheduler performs a match between SMBs and the outputs that share them using round-robin selection. Other selection schemes can be used instead.

The buffered crossbar has $\frac{N}{2}$ output-access schedulers, one for each pair of outputs. Multicast cells at the inputs have an N-bit multicast bitmap to indicate the destination of the multicast cells. Each bit of the bitmap is denoted as D_j, where $D_j = 1$ if output j is one of the cell destinations; otherwise $D_j = 0$. Each time a multicast copy is forwarded to the SMB for the cell's destination, the corresponding bit in the bitmap is reset. When all bits of a multicast bitmap are zero, the multicast cell is considered completely served. Call splitting is used by this switch to allow effective replication and to alleviate a possible head-of-line blocking.

A flow control mechanism is used to notify the inputs about which output replicates a multicast copy and to avoid buffer overflow. The flow control allows the inputs to send a copy of the multicast cell to the crossbar if there is at least one outstanding copy and an available SMB for the destined output. After all copies of the head-of-line multicast cell have been replicated, the input considers that cell served and starts the process with the cell behind.

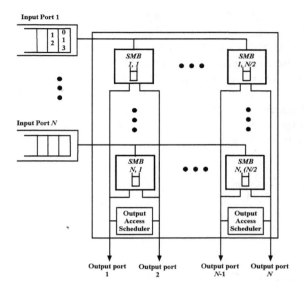

FIGURE 7.9

$N \times N$ O-SMCB switch with shared-memory crosspoints by outputs.

7.6 Performance Analysis of Shared-Memory Switches

This section discusses the performance of a shared-memory packet switch, such as the SMCB switch, through theoretical models. We focus on throughput and delay of the switch. This analysis consider the following assumptions:

- Packet arrivals are independent identically distributed Poisson processes with mean arrival rate λ_i for input port i, where $1 \leq i \leq N$.

- Packet lengths are exponentially distributed (i.e., service time at each output port is exponentially distributed with mean service rate μ_j, where $1 \leq j \leq N$).

These assumptions allow us to model the buffer management policies in shared-memory switch using queueing theory [93]. Here, we give an example of a probability-based approach in analyzing the throughput performance of a shared-memory switch — the mSMCB switch. We start with analyzing the blocking probability, P_b, at each input.

Because we assume each input independent of each other and uniformly distributed traffic, meaning all inputs have the same probability of receiving packets and the destination ports of each packet are also uniformly distributed, the throughput of the switch can be estimated as $1 - P_b$.

The blocking probability of a VOQ in the mSMCB switch with $k_s = 1$, denoted as P_b, is defined by two possible cases: 1) When the SMB is full with probability P_f. In this case, P_b is a function of the probability that a cell is forwarded to the corresponding output. The probability that there is a cell destined to this specific output is $\frac{1}{N}$. 2) When a given input contends with t inputs (where $0 \le t \le m-1$) for access to an available SMB, and the input is not granted because another input is matched. The probability that an input receives no grant is $\frac{t}{t+1}$. Considering these two cases, the blocking probability for the mSMCB switch is stated as

$$P_b = \frac{1}{N}P_f + \sum_{t=0}^{m-1} \binom{m-1}{t} \left(\frac{1}{N}\right)^t \left(1 - \frac{1}{N}\right)^{m-1-t} \frac{t}{t+1}(1 - P_f) \qquad (7.1)$$

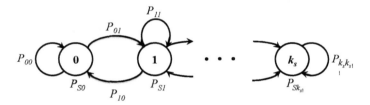

FIGURE 7.10
Diagram of the Markov chain of a shared crosspoint buffer of size k_s.

Figure 7.10 shows a Markov chain describing the occupancy of $SMB(q,j)$ in an mSMCB switch. P_{Sy} represents the state probability, where $0 \le y \le k_s$. P_{uv} is the transition probability from state u to state v. The probability that the SMB is full P_f is equivalent to the state probability P_{Sk_s}. For any k_s, P_{01} is defined by the product of the probability of input arrival $\rho_{i,j}$ and the matching probability between the inputs and the SMBs,

$$P_{01} = \rho_{i,j} \sum_{t=0}^{m-1} \binom{m-1}{t} \left(\frac{1}{N}\right)^t \left(1 - \frac{1}{N}\right)^{m-1-t} \frac{1}{t+1} \qquad (7.2)$$

The service probability $P_{service}$ is the probability that the output arbiter selects a nonempty $SMB(q,j)$ to forward a cell to the output. For $m = 2$, $P_{service} = \frac{2}{N}$ for any state. P_{10}, which occurs when input i has no requests and $SMB(q,j)$ is selected by the output arbiter, is defined by

$$P_{10} = (1 - \rho_{i,j})P_{service} \qquad (7.3)$$

The service probability $P_{service}$ is the probability that the output arbiter selects $SMB(q,j)$ to forward a cell, or $\frac{1}{N}$. The following balance equations are obtained when $k_s = 1$:

$$\begin{cases} P_{01}P_{S0} = P_{10}P_{S1} \\ P_{S0} + P_{S1} = 1 \end{cases}$$

$$(7.4)$$

The probability that the SMB is full is represented as

$$P_f = P_{S1} = \frac{P_{01}}{P_{01} + P_{10}} \tag{7.5}$$

7.7 Buffer Management for Shared-Memory Switches

Having shared memory for the switch fabric moves switching and queueing to the same physical location. This allows flexible access control to the shared memory and implementation of priority control and multicast services. The shared-memory switch architecture implements logical FIFO queues for each link to control access to the memory, and building queues in this memory requires exercising buffer management [50, 94]. The motivation of using buffer management is to answer questions like how to allocate the buffer space to different ports or flows to achieve higher throughput, memory utilization, and delay performance. We can consider three broad categories of buffer management schemes [9]: static threshold-based, push-out-based, and dynamic schemes. Many of the buffer management schemes find their roots from resource allocation and queueing theory [80, 93, 170]. We highlight a few featured schemes in each category below.

7.7.1 Static Threshold-Based schemes

Static threshold-based schemes allocate a fixed amount of buffer space to each input/output port in the shared memory [83, 87]. Based on how the threshold is decided, this type of scheme includes:

- Complete partitioning: The buffer space is partitioned among N input/output ports. This is the extreme case where there is no sharing among the input/output ports. This scheme may follow a rigid memory access scheme.

- Complete sharing: The buffer space is shared among all N input/output ports. Any accepted packet may be stored in any available space of the memory. This the opposite extreme case, where all input/output ports share the memory. This scheme provides maximum flexibility in memory access by the sharing ports. However, this flexibility could lead to monopolization of the memory by a single port when packets from this port come in at a higher rate in comparison to the traffic from other ports.

- Sharing with maximum queue length (SMXQ) [37]: This scheme imposes a limit on the number of buffers that can be allocated to each input/output port at any time to avoid memory monopolization by a single port. For an $N \times N$ switch, k_i is the size of buffer allocated to port i, M is the size of the memory, and α is defined as threshold parameter, where $\frac{1}{N} < \alpha < 1$,

$$k_i = \alpha M$$

Note that if $\alpha = \frac{1}{N}$, the policy becomes complete partitioning, and if $\alpha = 1$, the policy becomes complete sharing.

- Sharing with a minimum allocation (SMA) [87]: this scheme reserves a minimum number of buffers for each port. This ensures high-priority traffic from each port can always utilize at least the minimum reserved buffer space to avoid starvation.

- Sharing with a maximum queue and minimum allocation (SMQMA) [87]: This scheme integrates SMA and the previously introduced SMXQ, which sets an upper bound and lower bound of the buffer space for each port. The scheme is aimed to find a balance between avoiding starvation and maintaining flexibility of resource sharing.

7.7.2   Push-Out Schemes

Push-out schemes allow some of the packets that are already stored in the shared memory to be dropped when the memory becomes full to make space for higher priority packets. This approach is aimed at supporting multiple classes of traffic [34, 39, 182]. The decision of which packet (and from which queue) is dropped defines the different policies. A few examples of these policies are: drop-on-demand, complete sharing with virtual partition, and push-out with threshold. These policies are summarized below.

- Drop-on-demand [182]: This policy drops a packet from the longest queue to accommodate the new arriving packet when the buffer is full.

- Complete sharing with virtual partition (CSVP) [184]: This scheme virtually partitions the buffer space into N segments. When the buffer is not full, packets from all ports are accepted. When the buffer becomes full, if the occupancy of this class of packet is smaller than its allocation, it pushes out a packet to accept the incoming one.

- Push-out with threshold [39]: This scheme is similar to CSVP, but it implements the policy of dropping a packet from the longest queue to accommodate the new packet when the buffer is full.

7.7.3 Dynamic Schemes

As the traffic in the network changes with time, a switch has to be able to react to these traffic dynamics and change its buffer management policy for different arrival patterns. Some of the schemes introduced above incorporate a dynamic policy to react to the changing characteristics of the traffic [37, 169, 171]. We list some of these policies below.

- Adaptive control [169]: These schemes model the switching system as a dynamic control system. The measurement of traffic characteristics is considered the identification process and the policy of buffer allocation is considered the actuation process. The problem formulation becomes a change of system input (traffic arrivals) between time t and t_o.

- Dynamic threshold [37]: These schemes are similar to the threshold-based schemes discussed earlier. However, the thresholds of queue length of each port at a given time are set to be proportional to the amount of unused buffers in the switch. Different from the static threshold-based schemes, these schemes are formulated as a function of time t. If M is the total amount of memory, $Q(t)$ is the sum of all queue lengths, and α is the threshold parameter. The dynamic threshold scheme can be formulated as

$$T(t) = \alpha(M - Q(t))$$

The system will have a transient period to react to traffic load changes. By adjusting the threshold parameter α, the rate of system convergence can be controlled.

7.8 Exercises

1. For a switch with a single block of shared memory, 1 Gbps port rates and eight ports, what is the required memory speed to support 100% throughput of the switch assuming the ports are half duplex and uniform traffic if each cell has a size of 64 bytes?

2. Calculate the memory speed as in Exercise 1, but for full-duplex ports.

3. Calculate the memory speed as in Exercise 1, but for a switch that uses two blocks of memory, where each block is shared by $\frac{N}{2}$ output ports.

4. Calculate the memory speed as in Exercise 3, but for full duplex ports.

5. Assume that N is multiple of four. What is the speedup needed for unicast and multicast traffic if multicast packets are stored in unicast queues and the switch uses four blocks of memory, each block shared by $\frac{N}{4}$ ports?

6. Assume that the writing and reading process requires five memory access in a single-block shared-memory switch (Figure 7.1). Write an inequality that shows the relationship between the memory speed, port data rate, and the speedup required by this switch.

7. Indicate which packets would come out from the SMCB×2 switch in time slot $T + 3$ from the example in Section 7.3.

8. Indicate which packets would come out from the 2SMCB switch in time slot $T + 3$ from the example in Section 7.4.

9. List the advantages and disadvantages of the SMCB×2 switch.

10. List the advantages and disadvantages of the 2SMCB switch.

11. Discuss the implications of increasing m in the SMCB×m and mSMCB switches.

8

Internally Buffered Packet Switches

CONTENTS

Network traffic coming through a switch may face output contention. Blocking switches are susceptible to contention for internal links, such as Banyan switches. To avoid losing packets that lose contention for internal resources or for output ports, buffers may be allocated at the switching elements of the fabric rather than outside, such as in input- or output-queued switches. Switches with buffers in the fabric are called internally buffered. In these switches, memory may be dedicated to specific input-output interconnection or be shared by different interconnections. In this chapter, we focus on internally buffered nonblocking switches that have dedicated memory. Specifically, the considered switch fabric is a crossbar.

8.1   Introduction

Switches may include memory, organized as buffers, in the switch fabric to store packets that may lose output contention. The memory can be placed on different places of the fabric and that placement may determine the operation of the switch, and in turn, the performance of the switch. The switches considered in this chapter are internally buffered crossbars. The internal buffers of the switch are associated with the crosspoint elements of the crossbar and are called crosspoint buffers.

Buffered crossbars have been around for some time [177]. One of the initial concerns on buffered crossbars was that memory was expensive so it became of interest to design switches with small crosspoint buffers so as to reduce costs and make a buffered crossbar more affordable. With the advances in memory technology, the amount of memory that can be added to a buffered crossbar has significantly increased. However, other challenges, such as memory access speed, remain. In the following sections, we discuss several approaches of internally buffered switches.

8.2   Buffered-Crossbar Switch

Buffered crossbars, with N input ports and N output ports, have N^2 crosspoints. Each crosspoint is allocated a crosspoint buffer (CPB) to store packets from the input port that are to be forwarded to the output of the crosspoint. Therefore, a buffered crossbar has N^2 crosspoint buffers. Figure 8.1 shows a buffered-crossbar switch with N inputs and outputs.

Early buffered crossbar switches had a small number of ports as crosspoint buffers were expected to be as large as possible [22, 137]. For example, an early buffered crossbar switch set the number of ports to two (a 2×2 switch) to allow the largest possible crosspoint size, or 16 kbit at the time, to minimize packet losses [125].

As the memory technology matured, the implementation of buffered crossbar switches became more cost effective. Nowadays, it is easier to embed larger amounts of memory on chips [188].

In a buffered crossbar, packets are forwarded to the buffered crosspoint right after arriving in the input ports. A packet, among those buffered by the crosspoint buffers of an output port, is selected by an output arbiter for forwarding to the output port. The output arbiter considers all crosspoint buffers that have at least one stored packet and selects one crosspoint for forwarding a packet according to a selection policy. Therefore, the time and computation complexity of an output arbitration scheme may be in function of the number of crosspoint buffers (or N inputs) for an output.

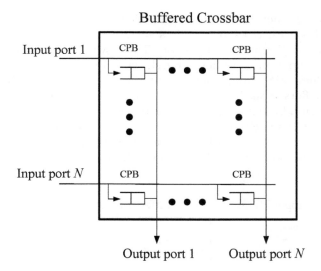

FIGURE 8.1

$N \times N$ buffered-crossbar switch.

Several arbitration schemes for buffered crossbar switches have been proposed [88, 125, 130, 188]. These works show that the packet loss ratio may be decreased by selecting an efficient arbitration scheme. However, because the amount of external memory that can be dedicated to crosspoint buffers may be larger than the internal one, allocating buffers at the inputs became of interest. The consideration of input buffers would decrease or eliminate packet losses. Buffered-crossbar switches with input buffers are called combined input-crosspoint buffered (CICB) packet switches.

8.3 Combined Input-Crosspoint Buffered (CICB) Packet Switch

Buffers at the input ports may be used to reduce the crosspoint buffer size. The inputs may be able to accommodate large amounts of memory at the input ports (outside the buffered crossbar).

8.3.1 CICB with First-In First-Out (FIFO) Input Buffers

Early versions of buffered-crossbar switches were an input and crosspoint buffering matrix with FIFO input buffers, or FIFO-CICB switch [188], and a CICB switch with input buffers and random selection policy at the output

[138]. It has been shown that the switch achieves high throughput. In addition, CICB switches with crosspoint buffers small enough to store no more than a single cell have been studied [73, 74]. These switches also use FIFO input buffers at the input ports, as the switch shown in Figure 8.2. The switches were shown to achieve 91% throughput despite suffering from the HoL blocking [80]. The FIFO buffers at the inputs limit the maximum throughput in that switch because the HoL blocking cannot be completely eliminated. Another example of a FIFO-CICB switch, with a different architecture approach, a multiple-plane architecture, is the cell-based tandem-crosspoint (TDXP) switch [131]. These switches show the need to remove the HoL blocking of FIFO-CICB switches.

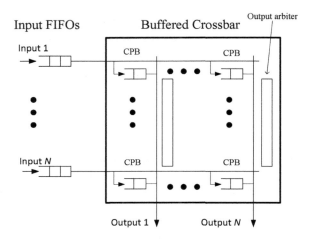

FIGURE 8.2
$N \times N$ buffered-crossbar switch with input FIFOs.

8.3.2 CICB with Virtual Output Queues at Inputs

The HOL blocking problem in FIFO-CICBs can be overcome by using virtual output queues (VOQs), where a VOQ is placed at an input to store packets or cells destined to a specific output. CICB switches with VOQs are denoted as VOQ-CICB (see Figure 8.3). However, for the sake of brevity, we refer to VOQ-CICB switches as CICB switches in the remainder of this chapter.

Consider a CICB switch with N inputs/outputs. In this switch model, there are N VOQs at each input. A VOQ at input i, where $0 \leq i \leq N$, that stores cells for output j, where $1 \leq j \leq N$, is denoted as $VOQ_{i,j}$. A crosspoint (CP) element in the CICB that connects input port i to output port j is denoted as $CP_{i,j}$. The buffer at $CP_{i,j}$ is denoted as $CPB_{i,j}$. The size of $CPB_{i,j}$, k, is indicated by the number of cells that can be stored. A credit-based flow-control mechanism indicates to input i whether $CPB_{i,j}$ has room

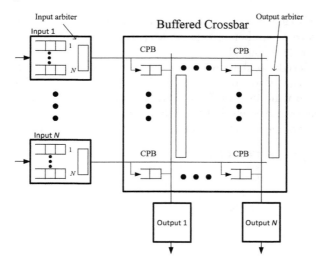

FIGURE 8.3
$N \times N$ buffered-crossbar switch.

available for a cell or not [139]. $VOQ_{i,j}$ is said to be eligible for selection if the VOQ is not empty and the corresponding $CPB_{i,j}$ at the buffered crossbar has room to store a cell.

In a switch with buffers, it is important to avoid losing cells by overflowing the receiving buffers. At the same time, restricting the forwarding may leave the buffer empty for some periods of time, underflowing the buffer. Both phenomena may reduce the achievable throughput of a CICB switch but a flow control mechanism may be used to avoid this throughput decrement. The flow control mechanism may send information between the crosspoint buffers and inputs to keep cells being forwarded without overflowing or underflowing the buffers. In a realistic implementation, those signals are propagated between the two components, and the flow control mechanism must consider the propagation times as well as the arbitration times. The round trip time (RTT) is defined as the sum of the delays of the input arbitration (IA), the transmission and propagation of a cell from an input to the crossbar ($d1$), the output arbitration (OA), and the transmission and propagation of the flow-control information back from the crossbar to the input ($d2$) [139]. Figure 8.4 shows an example of RTT for an input and a crosspoint buffer. The figure shows the transmission delays for $d1$ and $d2$, and arbitration times, IA and OA. Cell and bit alignments are included in the transmission times. The following condition must be satisfied for this switch to avoid buffer underflow:

$$\text{RTT} = d1 + \text{OA} + d2 + \text{IA} \leq k \qquad (8.1)$$

where k is the crosspoint buffer size (in time slots) which is equivalent to the number of cells that can be stored. In other words, the crosspoint buffer must be able to store a number of cells to keep the buffer busy (i.e., transmitting cells) during at least one RTT.

As the buffered crossbar switch may be physically placed far from the input ports in an actual implementation, RTTs may be nonnegligible. To support long round-trip time in a buffered-crossbar switch, the crosspoint-buffer size needs to be increased, such that up to RTT cells can be buffered. Nonnegligible round trip delays have been considered for practical implementations [3, 144].

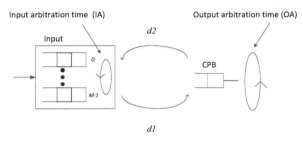

$d1$: Transmission delay from input i to crossbar
$d2$: Transmission delay from crossbar to input i
RTT $=$ IA $+ d1 +$ OA $+ d2$

FIGURE 8.4
Latencies in an RTT for avoiding buffer underflow.

8.3.3   Separating Arbitration in CICB Switches

CICB switches use selection time efficiently as input and output port selections are performed separately. For each input in a CICB switch, there is one input arbiter, which separately resolves input contention, i.e., a VOQ is selected to transfer a cell into the switch core. The arbitration of an input is independent from the arbitration of the other inputs, and from output arbitration. In a similar way, there is one output arbiter that independently handles output contention at each output. The output arbiter decides which crosspoint buffer is allowed to transfer a cell out of the switch core. Output arbitration is performed separately from the arbitration performed at the other outputs as well as from input arbitration. The input and output arbiters are coupled by the flow control mechanism [97]. Each crosspoint buffer and corresponding VOQ has an associated credit, which is used as a state flag of the crosspoint buffer (as being empty or occupied). During the input scheduling phase, the input scheduler at input i selects a non-empty $VOQ_{i,j}$ whose credit state is the buffer size (in number of cells). This credit decreases for each transmitted cell until reaching zero. The output arbiter at each input j selects a non-empty crosspoint buffer $CBP_{i,j}$. The credit at the input is increased for each cell selected by the output arbiter, as notified by the flow control. As an example of the strict timing for performing selection of a cell, a CICB switch with 40-

Gbps and 80-byte packets has 16 ns to perform input (or output) arbitration. For comparison, an input-buffered switch must perform matching, including input and output selection plus the time to exchange selection information between the input and output arbiters, in the same allowed time. It is then clear that timing in a CICB switch is more relaxed than that in an input-buffered switch.

8.3.4 Flow Control Mechanism

Using memory in the switch requires mechanisms to avoid overflowing and underflowing the memory with packets. Overflow would cause the loss of packets, which may require protocols in the transport layer to issue retransmissions. These retransmissions use bandwidth that may well be used for other packets. On the other hand, underflow occurs when an output may not be able to forward a packet out as packets are not yet stored in the crosspoint buffer while having packets available at the input. The CICB described in Section 8.3.2 uses a credit-based mechanism [96], where an input has N counters. Each time the input sends a cell to $CPB_{i,j}$, counter $C_{i,j}$ increases the count by one and each time a cell is transmitted out of the same CPB, the input is notified by the buffered crossbar and the counter is decreased by one. The maximum count is set to the size of the CPB to avoid overflow. The size of the CPB is set equal to the time it would take to send the output-arbiter decision back to the input, the time the input takes to make a queue selection, the time the new cell takes to reach the CPB, and the time the output arbiter may take to select the cell. This system is described in Equation 8.1.

There are other mechanisms to perform this flow control. One of them is backpressure [126], where instead of using a counter, an input uses an ON-OFF gate to allow a VOQ to forward a packet to the buffered crossbar or to stop the forwarding. In this mechanism, the crosspoint sends an OFF signal once it is full, and an ON signal once it is empty. However, this mechanism may not be suitable for CICB switches as the OFF signal may be issued too late to avoid overflow or the ON signal may be issued too late to avoid underflow as the propagation time must be considered.

8.4 Arbitration Schemes for CICB Switches

8.4.1 Oldest-Cell First (OCF) and Round-Robin Arbitrations

One of the advantages of having large buffers as VOQs in a switch is that weights can be assigned to them in different ways, and an input arbiter can perform VOQ selection based on those weights. Weight-based selection schemes can also be applied to CICB switches, at the input and output arbiters. The oldest cell first (OCF) selection for input arbitration and round-robin selection schemes for output arbitration have been considered for CICB switches [119, 86]. The OCF policy selects the packet (and the storing queue) with the longest waiting time. The combination of these selection schemes is denoted as OCF-RR. The performance of a CICB switch using these selection schemes and VOQs at the input ports showed that the crosspoint-buffer size can be small if the VOQs are provided with enough storage capacity.

8.4.2 Longest Queue First and Round-Robin Arbitration

The longest queue first (LQF) selection has been used in input-buffered switches to achieve high throughput under nonuniform traffic [113]. In a similar way, LQF may be applied to CICB switches [86]. As discussed before, LQF first selects the queue with the longest number of packets. Through a fluid model, it was proved that a CICB switch with the combination of LQF as input selection and RR as output selection (LQF-RR), with one-cell crosspoint buffers, achieves 100% throughput, where each input-output pair has a load of no more than $1/N$ (i.e., traffic with a uniform distribution) of a port capacity [45].

8.4.3 Most Critical Internal Buffer First

The most critical internal buffer first (MCBF) is another weighted-based arbitration scheme [114]. This scheme does not consider the state of a single crosspoint buffer, but rather the set of crosspoint buffers in the buffered crossbar. MCBF favors the least occupied buffer at the buffered crossbar to select which VOQ is serviced at an input, through the selection scheme named shortest internal buffer first (SBF), while the output side favors the most occupied internal buffer, named the longest internal buffer first (LBF).

In a switch with MCBF, the line of crosspoint buffers, denoted as $LCPB_i$, is defined as the set of all the internal buffers ($CPB_{i,j}$) at input i that store packets for all the outputs. NLB_i is the number of cells in $LCPB_i$. The column of crosspoint buffers $CCPB_i$ is defined as the set of all the internal buffers ($CPB_{i,j}$) that correspond to the same output j and receiving cells from all the inputs. NCB_j is the number of cells in $CCPB_j$. An additional schedule, with pointers in each arbiter, may be applied to break ties.

The input (SBF) and output (LBF) arbitration schemes work as follows:

- Input arbitration SBF scheme (selection of output j): For each input i, starting from the pointer position, select the first eligible VOQ corresponding to $\min_j NCB_j$ and send its HoL cell to the internal buffer $CPB_{i,j}$. Update the pointer position to $(j+1) \mod N$.

- Output arbitration LBF scheme (selection of input i): For each output j, starting from the pointer position, select the first $CPB_{i,j}$ corresponding to

$\max_i NCB_i$ and send its HoL cell to the output. Update the pointer position to $(i+1) \mod N$.

Figure 8.5 shows an example of VOQ selection by an input arbiter in a 4×4 switch, where $0 \le i, j \le 3$. The state of the switch is represented as a bi-dimensional matrix, where rows represent the inputs and columns represent the outputs. Figure 8.5(a) shows the state of CPBs as 1 for those that have a cell in it and 0 for those CPBs that are empty. Figure 8.5(b) shows the column occupancy as seen by each input (matrix row). For example, Input 1, the second row from the top, may select to forward a cell to $CPB_{0,2}$ and $CPB_{0,3}$. Therefore, Input 0 selects a cell from either $VOQ_{0,2}$ or $VOQ_{0,3}$. The column occupancies are then one and two cells for Outputs 3 and 2, respectively. Therefore, SBF selects $j = 3$ or $VOQ_{0,3}$. Figure 8.5(c) shows the final selection result, where selected CPBs are marked as X.

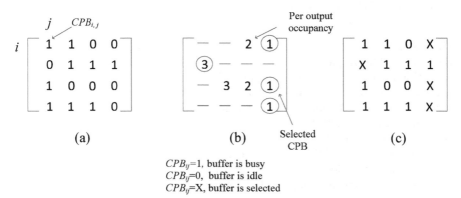

(a) (b) (c)

Selected CPB

$CPB_{ij}=1$, buffer is busy
$CPB_{ij}=0$, buffer is idle
$CPB_{ij}=X$, buffer is selected

FIGURE 8.5
Example of SBF in a 4x4 switch.

Figure 8.6 shows an example of the selection process of a CPB that LBF performs in the same 4×4 switch. Figure 8.6(a) shows the state of CPBs as busy and idle crosspoints. Figure 8.6(b) shows the input occupancy seen by the output arbiters per CPB. For example, arbiter at Output 2 selects the row with the largest number of cells, or row representing Input 1. The other output arbiters perform their selection in a similar way. Figure 8.5(c) shows that $CPB_{2,0}$ and $CPB_{1,1}$ to $CPB_{1,3}$ are selected by the outputs.

The performance of MCBF is similar to those of LQF-RR and OCF-OCF, which are combined arbitration schemes. MCBF also uses pointers and a supporting scheme to break ties.

8.4.4 Round-Robin (RR) Arbitration

Round-robin (RR) selection has been used in input-buffered packet switches to provide high throughput under uniform traffic [110]. Arbitration based on

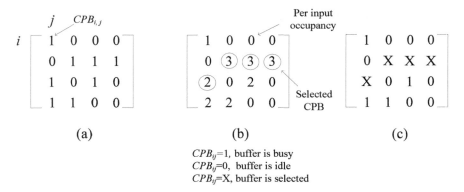

FIGURE 8.6
Example of LBF in a 4x4 switch.

RR is attractive because it provides a high level of fairness and it is simple to implement. The implementation feature is often associated with arbitration resolution at high speeds. As in input-buffered switches, round-robin selection has been used in CICB switches [86, 139]. An example is the CIXB-1 switch, which was used to demonstrate that using crosspoint buffers of the minimum size, one cell, and round-robin as input and output selection schemes 100% throughput under uniform traffic is achievable [139]. In this switch, the input arbiter selects a noninhibited and nonempty VOQ. The inhibition of VOQs is determined by the flow control mechanism. A VOQ is said to be inhibited if the crosspoint corresponding to that input and output is full.

The output arbiter selects a crosspoint buffer from those eligible using RR. Similarly, a crosspoint buffer is eligible if it has a packet stored in it. After an arbiter selects a VOQ (at an input) or a crosspoint buffer (at an output), the pointer is moved to one position beyond the selected queue.

The CIXB-1 switch has several attractive properties: it uses a simple arbitration scheme and the time allowed for arbitration is more relaxed than the time allowed to perform matching in an input-buffered switch, as the switch does not incur interchip or interboard delays. The relaxed timing allows to handle fast link rates. The arbitration complexity of RR is O(N), which can be reduced to O(log N) with a suitable encoding approach. This switch has been shown to provide 100% throughput under uniform traffic with Bernoulli and bursty arrivals. CIXB-1 showed that CICB switches can deliver higher throughput than input-queued switches using similar arbitration schemes.

Compared to the performance of an IB switch running *i*SLIP [111] as the port-matching algorithm, CIXB-1 achieves lower queueing delay under uniform traffic and higher throughput under nonuniform traffic. However, as actual traffic may present nonuniform distributions, it is necessary to provide arbitration schemes that also provide 100% throughput under nonuniform traffic patterns.

8.4.5 Round-Robin with Adaptable Frame Size (RR-AF) Arbitration Scheme

Weight-based schemes are known to perform well for nonuniform traffic patterns. These schemes perform comparisons among all contending queues, which can be a large number. Those comparisons take time and resources. Moreover, queue-occupancy-based selection may starve some queues for a long time to provide more service to the congested ones. On the other hand, round-robin schemes have been shown to provide fairness and implementation simplicity, as no comparisons are needed among queues and are well suited to provide high performance under uniform traffic [111]. However, round-robin selection may not be suitable for switching nonuniform traffic. For example, it has been shown that a switch using RR needs large crosspoint buffers to provide high throughput under admissible unbalanced traffic [151], where the unbalanced traffic model is a nonuniform traffic pattern [139]. This large buffer can make the implementation of a switch costly.

Frame-based matching has been shown to have improved switching performance under different traffic scenarios [16]. However, how to set the frame size is a complex issue. The round-robin with adaptable-size frame (RR-AF) scheme was proposed to avoid assigning frame sizes arbitrarily [149]. RR-AF is based on the amount of service that a buffer gets. Each time a VOQ (or a CPB at an output) is selected by the arbiter, the VOQ gets the right to forward a frame, where a frame is formed by one or more cells. Each cell of a frame is dispatched in one time slot. The RR-AF scheme can achieve nearly 100% throughput under admissible uniform and nonuniform traffic. In RR-AF, each VOQ (and CPB) has two counters: a frame-size counter, $FSC_{i,j}(t)$, and a current service counter, $CSC_{i,j}(t)$. The value of $FSC_{i,j}(t)$, —$FSC_{i,j}(t)$—, indicates the frame size; that is, the maximum number of cells that $VOQ_{i,j}$ is allowed to send in back-to-back time slots to the buffered crossbar, one cell per time slot. The initial value of —$FSC_{i,j}(t)$— is one cell (i.e., its minimum value). Here, —$FSC_{i,j}(t)$— can be as large as needed, although practical results have shown that its value remains rather small. $CSC_{i,j}(t)$ counts the number of serviced cells at time slot t in a frame corresponding to a VOQ, where the frame size is indicated by FSC, in a regressive fashion. A regressive-fashion count is used in CSC as CSC only considers FSC at the end of a serviced frame. The initial value of $CSC_{i,j}(t)$, —$CSC_{i,j}(t)$—, is one cell (i.e., its minimum value). The input arbitration process works as follows. An input arbiter selects an eligible $VOQ_{i,j'}$ in round-robin fashion, starting from the pointer position, j. For the selected $VOQ_{i,j'}$, if $|CSC_{i,j'}(t)| > 1$, $|CSC_{i,j'}(t+1)| = |CSC_{i,j'}(t)| - 1$, and the input pointer remains at $VOQ_{i,j'}$, so that this VOQ has a higher priority for service in the next time slot and the frame transmission continues. If $|CSC_{i,j'}(t+1)| = 1$, the input pointer is updated to $(j' + 1) \mod N$, $|FSC_{i',j'}(t)|$ is increased by f cells, and $|CSC_{i,j'}(t)| = |FSC_{i,j'}(t)|$. For any other $VOQ_{i,h}$, where $h \neq j'$, which is empty or inhibited by the flow-control mechanism, the pointer is positioned between the pointed $VOQ_{i,j}$ and the selected $VOQ_{i,j'}$: if $|FSC_{i,h}(t)| > 1$, $|FSC_{i,h}(t+1)| = |FSC_{i,h}(t)| - 1$. If there exist one or more VOQs that fit the description of $VOQ_{i,h}$ at a given time slot, it is said that those VOQs missed a service opportunity at that time slot. The increment of the frame size, done by f cells, is performed each time the previous complete frame of a VOQ has been serviced. The following pseudocode describes the input arbitration scheme, as seen at an input:

- At time slot t, starting from the pointer position j, find the nearest eligible $VOQ_{i,j'}$ in a round-robin fashion among eligible VOQs.

- Send the HoL cell from $VOQ_{i,j'}$ to $CPB_{i,j'}$ at time slot $t + 1$.

  - If $|CSC_{i,j'}(t)| > 1$ then $|CSC_{i,j'}(t+1)| = |CSC_{i,j'}(t)| - 1$, the pointer remains pointing to j'.

  - Else $|FSC_{i,j'}(t+1)| = |FSC_{i,j'}(t)| + f$, $|CSC_{i,j'}(t+1)| = |FSC_{i,j'}(t+1)|$, the pointer is moved to $(j' + 1)$ module N.

- For $VOQ_{i,h}$, where $j \leq h < j'$ for $j < j'$, or $0 \leq h < j'$ and $j \leq h \leq N-1$ for $j > j'$:

  - $|FSC_{i,h}(t+1)| = |FSC_{i,h}(t)| - 1$.

- Continue to the next time slot.

Here, $j < j'$ means that the arbiter considers j' before j in a round-robin order. It should be noted that f may be a constant or a function. In the original approach, f is set equal to N. The value of f affects the performance of RR-AF in different traffic scenarios. Note that if $f = 0$, RR-AF becomes RR. Figure 8.7 shows an example of the adjustment of $FSC_{i,j}$ in an input of a 4×4 switch. In this example, $VOQ_{i,2}$ and $VOQ_{i,3}$ store one and three cells, respectively (Figure 8.7(a) shows), and no VOQ is inhibited by the flow-control mechanism. At time slot t, the pointer of RR-AF points to $VOQ_{i,0}$. During this time slot, the input arbiter selects $VOQ_{i,2}$ to send a cell to the buffered crossbar. Then $VOQ_{i,0}$ and $VOQ_{i,1}$ miss an opportunity to send cells as they are empty and their FSCs are decreased by one at the end of the time slot. Note that $VOQ_{i,0}$ and $VOQ_{i,1}$ are considered to be $VOQ_{i,h}$ for this time slot as described in RR-AF. Table 8.1 shows the evolution of the FSC values for each VOQ during six time slots. In the next time slot, $t + 1$, $VOQ_{i,2}$ is served, and it becomes empty. As the pointer points to this VOQ, $FSC_{i,2}$ is decreased to 1 in the next time slot. Therefore, the arbiter selects $VOQ_{i,3}$ at time slot $t + 2$ as the next VOQ to receive service. Then the pointer is moved to $VOQ_{i,3}$. At time slot $t + 3$, $VOQ_{i,3}$ is again selected. Since the last frame cell of $VOQ_{i,3}$ is selected, $FSC_{i,3}$ is updated to $2 + N = 6$. However, since there are no more cells in this VOQ, $FSC_{i,3}$ decreases by one in the subsequent time slots. In this table, a dash in time slots $t + 4$ and $t + 5$ means that no j is selected. Figure 8.7(b) shows the order in which cells are served.

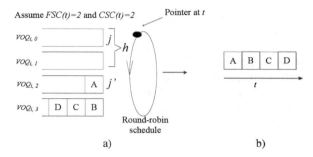

FIGURE 8.7
Example of the operation of RR-AF arbitration.

TABLE 8.1
Values in different time slots of frame counters in RR-AF.

FSC/j	Time Slots					
	t	$t+1$	$t+2$	$t+3$	$t+4$	$t+5$
$FSC_{i,0}$	2	1	1	1	1	1
$FSC_{i,1}$	2	1	1	1	1	1
$FSC_{i,2}$	2	2	1	1	1	1
$FSC_{i,3}$	2	2	2	6	5	4
Selected j	2	3	3	-	-	-

RR-AF delivers 100% throughput under uniform traffic. It also has an average cell delay close to that of an output buffered switch, as RR selection does. The efficiency of RR-AF is more noticeable under unbalanced traffic, which is a nonuniform traffic pattern, as RR-AF with a CPB size of one cell achieves higher throughput than RR.

8.5 Switching Internal Variable-Length Packets

The switches discussed so far perform internal transmission of fixed-size packets, or cells. Instead, original variable-length packets can be switches. In this case, segmentation of packets and re-assembly are not needed. Using cells internally in a switch requires that padding bytes be transferred when a packet length is not evenly divisible by the cell length. In the worst case, there could be cells with a size of S bytes and an incoming packet with the length of $S + 1$ bytes. To transmit this packet, two time slots would be needed. Therefore, to maintain wire speed transmission, an internal speedup of $2S/S + 1$, which is approximately two, is needed if no sophisticated segmentation is used

[164]. A segmentation mechanism, where a segment contains data from different packets, may show additional advantages [90]. In this way, padding bytes and speedup may not be needed for a CICB to keep up with wire speeds. A CICB switch using parallel polling was used to switch variable-length packets [187]. This switch used a round-robin order for polling VOQs and CPBs at the inputs and outputs, respectively. Polling is an alternative to arbitration. A CICB switching variable-length packets has been proposed [91]. This switch also uses round-robin selection for input and output arbitration.

8.6 Exercises

1. Investigate and discuss why the first buffered-crossbar switches were initially designed with a small number of ports.

2. Considering that input and output arbitration of a CICB switch takes one time slot, and a crosspoint buffer stores 10 time slots worth of packets, how long a time can be used for sending packets from inputs to the crosspoint buffers (i.e., $d1$) if $d1$ and $d2$ take the same time budget?

3. Describe how CICB switches avoid overflowing crosspoint buffers.

4. Describe how CICB switches avoid underflowing crosspoint buffers.

5. Consider that a crosspoint buffer in a CICB can store 20 time slots worth of packets. What is the highest achievable throughput of the switch should RTT consume 25 time slots?

6. Explain why input and output arbitration are *separated* in a CICB switch.

7. List the advantages and disadvantages of weighted arbitration schemes in buffered crossbar switches.

8. List the advantages and disadvantages of round-robin arbitration schemes in buffered crossbar switches.

9. Discuss the advantages and disadvantages of considering variable-length packet switching in buffered-crossbar switches.

10. Discuss how frame-based arbitration may increase the performance of a CICB switch.

9

Load-Balancing Switches

CONTENTS

Load-balancing packet switches have the objective to simplify switching operations by distributing the traffic coming into a switch before switching towards the destined output takes place. It is known that, in general, the required efficiency of a scheduler of a switch is less demanding under evenly distributed traffic as this distribution requires a similar level of service for all ports. This chapter presents some of these switches.

9.1 Introduction

Balancing the load of a network equalizes the required service for every portion of the load. In packet switches, round-robin schemes have been used to effectively switch uniform traffic. Therefore, there has been recent interest in balancing the traffic loads passing through a switch or router to lower the complexity of packet-scheduling schemes. With the increasing data rates in the

Internet and the variety of patterns that new network applications generate in the network, load-balancing schemes seem to be an attractive alternative.

This chapter introduces several load-balancing (also called load-balanced) packet switches. In general, the idea of load-balancing switches is to use one stage of the switch for balancing the incoming load and the other to complete switching packets to their destined outputs. In this chapter, we continue to focus on cell-switching schemes, where cells are segments of variable-length Internet packets.

9.2 Load Decomposition in Packet Switches

As presented in Chapter 5, traffic incoming to a switch and the configurations of a crossbar switch fabric can both be represented as a two-dimensional matrix. Therefore, as long as the traffic and its distribution can be profiled, this traffic can be decomposed into multiple permutation matrices, which are similar to identity matrices, that indicate the configurations needed by the switch fabric [25, 100].

Load decomposition for packet switching can show the relationship between switch configuration and load balancing. Load decomposition shows that different load patterns may be expressed in terms of permutation matrices. The set of permutation matrices may indicate that there is a uniformity component in the incoming traffic and an even distribution of it may help to simplify the required switch configurations for the switch to achieve an optimum performance. Experience from earlier works in packet switches using uniform traffic shows that round-robin and other scheduling schemes with low complexity are able to switch 100% of the traffic. Therefore, dependence on the scheduling scheme can be minimized with load balancing.

Several schemes to adopt this method were focused on Clos-network switches [25], which are complex to configure as they are multiple-stage switches. For single-stage switches, the decomposition method may be less complex. Decomposition of traffic patterns can be performed by following the Birkhoff and von Neumann methods [18, 176].

In this switch, the connection patterns of the crossbar can be represented by a combination of permutation matrices. This switch is called the Birkhoff–von Neumann (BVN) switch as it uses two algorithms for the traffic decomposition method [18, 176].

The calculation of the configuration matrices includes two algorithms:

- **Algorithm 1:** Construct a doubly stochastic matrix from the input traffic matrix.

- **Algorithm 2:** Find the permutation matrices and the coefficients forming the doubly stochastic matrix generated from Algorithm 1.

Let us define R as the traffic (rate) matrix, where $r_{i,j}$ is the rate assigned to the traffic from input i to output j for an $N \times N$ input-buffered crossbar switch. The following definition applies:

Doubly substochastic matrix: Assume a nonnegative matrix, $R = (r_{i,j})$, where $1 \leq i, j \leq N$, R is known to be a doubly substochastic matrix if it satisfies the following two conditions:

$$\sum_{i=1}^{N} r_{i,j} \leq 1, \forall j \qquad (9.1)$$

and

$$\sum_{j=1}^{N} r_{i,j} \leq 1, \forall i \qquad (9.2)$$

The above two conditions also indicate traffic admissibility. Before proceeding to obtain the permutation matrices, the doubly substochastic matrix is converted into a doubly stochastic one. The process of constructing a doubly stochastic matrix given a substochastic traffic (rate) matrix is described in Algorithm 1.

Algorithm 1: Construct a stochastic matrix

1. If the sum of all the elements in R is less than N, then there is an element (i, j) such that $\sum_{n} r_{i,n} < 1$ and $\sum_{m} r_{m,j} < 1$.

2. Let $e = 1 - max[\sum_{n} r_{i,n}, \sum_{m} r_{m,j}]$. Add e to the element in the $(i,j)^{th}$ position to obtain a new matrix $\tilde{R}$.

3. Repeat this procedure until the sum of all the elements is equal to N.

After obtaining $\tilde{R}$, matrix decomposition follows:

Algorithm 2: Matrix decomposition

1. For a doubly stochastic matrix $\tilde{R}$, let $(i_1, i_2, \cdots, i_N)$ be a permutation of $(1, 2, \cdots, N)$ such that $\Pi_{k=1}^{N} \tilde{r}_{k,i_k} > 0$.

2. Let P_1 be the permutation matrix corresponding to $(i_1, i_2, \cdots, i_N)$, and $\phi_1 = min_{1 \leq k \leq N}[r_{k,i_k}]$. Define the matrix R_1 by $R_1 = \tilde{R} - \phi_1 P_1$.

3. If $\phi_1 = 1$, then $R_1 e = \tilde{R} e - P_1 e = \mathbf{0}$, where $\mathbf{0}$ is the column vector with all its elements being zero.

4. If $\phi_1 < 1$, then the matrix $\frac{1}{1-\phi_1} R_1$ is doubly stochastic and we can repeat Step 1.

The following is a simple example of how a matrix can be decomposed into a linear combination of permutation matrices.

Example 1. A matrix

$$R = \begin{bmatrix} 1/4 & 1/4 & 1/4 & 1/4 \\ 1/4 & 1/4 & 1/4 & 1/4 \\ 1/4 & 1/4 & 1/4 & 1/4 \\ 1/4 & 1/4 & 1/4 & 1/4 \end{bmatrix}$$

is 1/4 sum of the following permutation matrices:

$$\begin{bmatrix} 1 & 0 & 0 & 0 \\ 0 & 1 & 0 & 0 \\ 0 & 0 & 1 & 0 \\ 0 & 0 & 0 & 1 \end{bmatrix}, \begin{bmatrix} 0 & 0 & 0 & 1 \\ 1 & 0 & 0 & 0 \\ 0 & 1 & 0 & 0 \\ 0 & 0 & 1 & 0 \end{bmatrix}, \begin{bmatrix} 0 & 0 & 1 & 0 \\ 0 & 0 & 0 & 1 \\ 1 & 0 & 0 & 0 \\ 0 & 1 & 0 & 0 \end{bmatrix}, \begin{bmatrix} 0 & 1 & 0 & 0 \\ 0 & 0 & 1 & 0 \\ 0 & 0 & 0 & 1 \\ 1 & 0 & 0 & 0 \end{bmatrix}$$

$$R = \frac{1}{4}\begin{bmatrix} 1 & 0 & 0 & 0 \\ 0 & 1 & 0 & 0 \\ 0 & 0 & 1 & 0 \\ 0 & 0 & 0 & 1 \end{bmatrix} + \frac{1}{4}\begin{bmatrix} 0 & 0 & 0 & 1 \\ 1 & 0 & 0 & 0 \\ 0 & 1 & 0 & 0 \\ 0 & 0 & 1 & 0 \end{bmatrix} + \frac{1}{4}\begin{bmatrix} 0 & 0 & 1 & 0 \\ 0 & 0 & 0 & 1 \\ 1 & 0 & 0 & 0 \\ 0 & 1 & 0 & 0 \end{bmatrix} \frac{1}{4}\begin{bmatrix} 0 & 1 & 0 & 0 \\ 0 & 0 & 1 & 0 \\ 0 & 0 & 0 & 1 \\ 1 & 0 & 0 & 0 \end{bmatrix}$$

These examples produce four different permutation matrices, each corresponding to the configuration of the switch fabric to interconnect inputs to outputs. By applying these found configurations in the ratio as the coefficients show, the traffic is fully served. Let us consider a traffic matrix where the rates are not as uniform as in the example above.

Example 2. Consider the following traffic (rate) matrix, R, as follows:

$$R = \begin{bmatrix} 0.1 & 0 & 0.13 & 0.4 \\ 0.11 & 0.07 & 0.6 & 0.05 \\ 0 & 0.32 & 0.12 & 0.23 \\ 0.51 & 0.23 & 0.04 & 0.12 \end{bmatrix}$$

Observe that the sum of each row and column is smaller than 1. This satisfies the doubly substochastic matrix conditions. We can apply Algorithm 1 to find elements for $\tilde{R}$. The maximum value of the sum of each row and each column is 0.9. So e is calculated as 0.1. Add this to (1,1), (1,3), (2,2), (3,2), (3,3), (3,4), and (4,4) and recalculate e. Do this iteratively until the sums of all rows and columns become 1. A doubly stochastic matrix $\tilde{R}$ in this example is then

$$\tilde{R} = \begin{bmatrix} 0.38 & 0 & 0.22 & 0.4 \\ 0.11 & 0.24 & 0.6 & 0.05 \\ 0 & 0.53 & 0.14 & 0.33 \\ 0.51 & 0.23 & 0.04 & 0.22 \end{bmatrix}$$

After applying Algorithm 2, $\tilde{R}$ becomes

$$\tilde{R} = 0.14 \begin{bmatrix} 1 & 0 & 0 & 0 \\ 0 & 1 & 0 & 0 \\ 0 & 0 & 1 & 0 \\ 0 & 0 & 0 & 1 \end{bmatrix} + 0.04 \begin{bmatrix} 1 & 0 & 0 & 0 \\ 0 & 0 & 0 & 1 \\ 0 & 1 & 0 & 0 \\ 0 & 0 & 1 & 0 \end{bmatrix}$$

$$+ 0.04 \begin{bmatrix} 0 & 0 & 0 & 1 \\ 0 & 0 & 1 & 0 \\ 0 & 1 & 0 & 0 \\ 1 & 0 & 0 & 0 \end{bmatrix} + 0.08 \begin{bmatrix} 1 & 0 & 0 & 0 \\ 0 & 0 & 1 & 0 \\ 0 & 1 & 0 & 0 \\ 0 & 0 & 0 & 1 \end{bmatrix}$$

$$+ 0.01 \begin{bmatrix} 0 & 0 & 1 & 0 \\ 0 & 0 & 0 & 1 \\ 0 & 1 & 0 & 0 \\ 1 & 0 & 0 & 0 \end{bmatrix} + 0.12 \begin{bmatrix} 1 & 0 & 0 & 0 \\ 0 & 0 & 1 & 0 \\ 0 & 0 & 0 & 1 \\ 0 & 1 & 0 & 0 \end{bmatrix}$$

$$+ 0.11 \begin{bmatrix} 0 & 0 & 1 & 0 \\ 1 & 0 & 0 & 0 \\ 0 & 0 & 0 & 1 \\ 0 & 1 & 0 & 0 \end{bmatrix} + 0.1 \begin{bmatrix} 0 & 0 & 1 & 0 \\ 0 & 1 & 0 & 0 \\ 0 & 0 & 0 & 1 \\ 1 & 0 & 0 & 0 \end{bmatrix}$$

This example produces six disjoint permutation matrices and a set of co-efficients. As before, by applying the configurations to the switch fabric as indicated by the permutation matrices, the incoming traffic would be effectively switched to the outputs.

However, in practical scenarios, the traffic distribution may only be known after long periods of time, and yet the traffic distribution may change over time.

9.3 Load-Balanced Birkhoff–von Neumann Switches

The Load-Balanced BVN (LB-BVN) switch may overcome the requirement of knowing the traffic distribution before the traffic arrives to the switch. In this switch, traffic is evenly distributed among the different inputs of the switch. This means that the switch uses a load-balancing stage and a switching stage.

Figure 9.1 shows the architecture of the LB-BVN switch. It has N input and output ports, a load balancer, and a switch that forwards cells to their destined output, called the BVN switch. The load-balancing stage and the BVN switch are implemented as crossbars using periodic permutations or configurations.

The first and second stages of the switch perform periodical preestablished connections. The two stages are agnostic to the incoming traffic pattern and to packet arrivals, and the switch fabric interconnects Input i to Output j

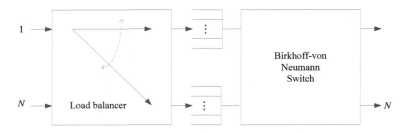

FIGURE 9.1
$N \times N$ Load-balanced Birkhoff–von Neumann switch.

where $i = (j + t \mod N) + 1$, where $1 \le i$, $j \le N$, and $t \ge 0$. Figure 9.2 shows an example of the configurations of the first and second stages of a 4×4 switch. The example shows four possible configurations. These four configurations are disjoint, as each input interconnects to a different output in each configuration.

The throughput achieved by this switch is 100% under admissible traffic. Note that no scheduling is needed in this switch as the configuration of the switch is cyclical and fixed, independently of the traffic pattern. Because the configuration of the switch follows a fixed pattern, the complexity is $O(1)$.

However, the switch suffers from delivering cells in out-of-sequence as a product of the load-balancing stage. To show this, let us consider an example in a 4×4 LB-BVN switch using the configurations Figure 9.2 shows, as order followed from (a) to (d). Let us also consider that three of the inputs receive the following traffic: Inputs 1 receives four cells for Output 4, Input 3 receives four cells for Output 1, and Input 4 receives four cells for Output 2. Each cell is labeled as X_y, where X is the destination port number and y is the order in which the cell was received. Here, **a** is the cell arriving first, cell **b** is the second, and so on. The **a** cells in each of the inputs receiving traffic arrive at time slot t, and so on, until the **d** cells arrive at time slot $t + 3$. Because the load-balancing stage forwards a cell towards the intermediate queues at arrival, the cells arrive at the indicated time slot and in the intermediate queue as indicated by the order in which Figure 9.3 shows. It should be noted that the intermediate queues are virtual output queues (VOQs), so the figure shows the order and times in which the cells arrive. After considering the arrivals in the intermediate outputs, the position of the cells in their VOQs, and the configuration of the BVN switch, the order and times in which the cells are forwarded to their destined outputs are as the figure shows in the right side of the figure. Note that cell 1c arrives at the output before 1b, 2c is also out of order, and so are 4b and 4d.

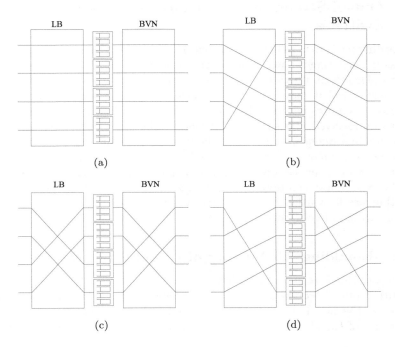

FIGURE 9.2
Example of configurations of the LB and BVN stages of a 4×4 LB-BVN switch.

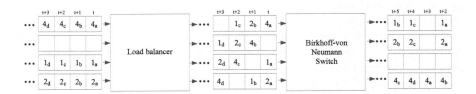

FIGURE 9.3
Example of switching cells through the LB-BVN switch and the order in which they exit the switch.

9.4   Out-of-Sequence Forwarding in Load-Balanced Switches

Forwarding packets in out-of-sequence from the output of a switch is not desirable as it may trigger congestion control schemes to modify the transmission of data. Therefore, providing measures to sort out-of-sequence cells is costly in resources and it may affect the performance of the switch. Sorting cells, or packets, requires additional buffers at the output line cards to wait for all cells to be received and sorted out without affecting the forwarding of cells that come in sequence. In addition, the waiting time for cells being late slows down the performance of the switch and it may affect other traffic.

This out-of-sequence problem has attracted the proposal of several schemes as solutions. These solutions are categorized as: a) bounding the number of cells that are forwarded to their outputs out-of-sequence, and b) preventing the forwarding of cells out-of-sequence.

9.4.1   Bounding the Number of Out-of-Sequence Cells

Schemes in this category aim to limit the number of cells that can be sent to the output out-of-sequence. Cells are re-assembled at the output before being sent out of the switch. Having a limit on the number of cells, it also sets a limit on the size of the re-assembly buffer. Some basic schemes are the following.

9.4.1.1   First-Come First-Serve (FCFS)

This scheme uses a flow splitter and load-balancing buffers at the front of the load-balancing stage, time stamps, and a jitter control mechanism placed between the load-balancer switch and the VOQs in the middle stage of the switch [27]. The flow splitter and the load-balancing buffers are used to spread arriving cells according to the order they arrive rather than by the time they arrive. Otherwise, there is no guarantee on the size of the re-sequencing and output buffer to prevent packet losses [84]. In this switch, arriving cells are flow split. Once they are forwarded through the load-balancing stage to the VOQs, early cells are delayed by the jitter control mechanism so older cells that are expected to arrive first to the output can catch up. Figure 9.4 shows the architecture of an LB-BVN switch adopting FCFS. However, the use of the jitter control mechanism may impact the performance of the switch.

9.4.1.2   Earlier Deadline First (EDF)

This scheme is also based on the two-stage switch used by the FCFS schemes, using a flow splitter and load-balancing buffers in front of the inputs [27]. In addition, EDF requires a reference OQ switch using FCFS at the output

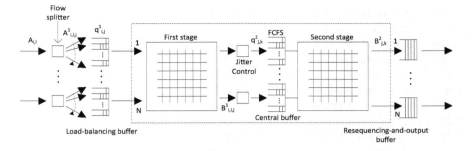

FIGURE 9.4
Switch architecture for application of FCFS.

queues to determine the departure time of a cell should the cell be switched by such a switch. The OQ departure time of a cell is used as a time stamp to determine the urgency of the cell. Figure 9.5 shows the architecture of an LB-BVN switch adopting EDF. However, monitoring the reference OQ switch

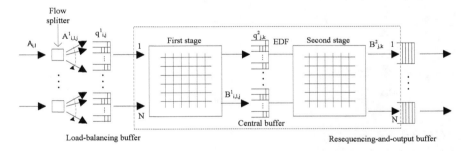

FIGURE 9.5
Switch architecture for the application of EDF.

and scheduling packet according to their time stamps is complex. Furthermore, the number of time stamps is large and keeping them requires resources and a large number of bits to avoid wraparound values in practical implementations.

9.4.2 Preventing Out-of-Sequence Forwarding

9.4.2.1 Full Frame First (FFF) [92]

This scheme uses three-dimensional queues (3DQs) in-between the first and second stages of the switch. These 3DQs are an extension of VOQs used to avoid combining cells from different flows (here also defined as cells going from an input and destined to an output of the switch, $F_{i,j}$) in the same queue. In other words, a 3DQ is a virtual input-output queue. That is, there is a queue at an internal input (II) that queues cells from external input (EI)

i, EI_k, to external output (EO) k, EO_k. Figure 9.6 shows the 3DQs in the switch architecture. In the figure, VOQ_1 is the load-balancing queues and the 3DQs are virtual EI and EO queues. As before, cells of a flow are sent to different load-balancing queues, independently of the arrival time but in the order in which they arrive. After that, cells of a flow are dispersed through the 3DQs by the load-balancing stage and since no flows are combined, they may be dispatched toward the outputs without being delayed by cells of other flows and, therefore, without causing out-of-order forwarding. In this scheme, a frame is the set of up to N cells that can be forwarded to the output if the queues are selected in a round-robin schedule (one per time slot and in consecutive time slots) and the first cell of the frame is the next in order to be served. To identify the order, cells are time stamped. If a frame can forward up to N cells in order, the frame is determined full, and nonfull otherwise. This definition does not require N cells, but all cells from the pointer position and up to N queues to serve cells in order. The scheme then gives priority to full frames over nonfull frames.

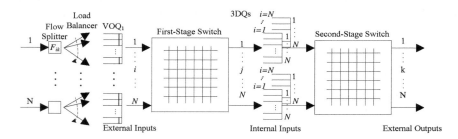

FIGURE 9.6
Switch architecture for the application of FFF.

Because round-robin schedule is used, there is a full-frame pointer for each EO, $p_{ff}(k)$, to indicate which external output k must be served in the round-robin order, and for each $p_{ff}(k)$ there is a frame pointer (or full-frame II pointer), $p_{II\ k}$, to indicate which VOQ_1 stores the head of the frame cell. In addition, each EO also has a nonfull frame pointer, $p_{nff}(k)$, to indicate which EI is next in the order for nonfull frame service. The pointers are updated to the one position beyond the last served after service (either full or nonfull).

The algorithm is as follows:

1. Determine which of the frames $F_{i,k}$ is full, where $i \in \{1, \ldots, N\}$.

2. Starting at $p_{ff}(k)$, find the first full frame. If the first full frame arrived from external input i_{ff}, then $p_{ff} \leftarrow i_{ff} + 1$, modulo N. If there is no full frame, $p_{ff}(k)$ doesn't change.

3. If there is no full frame, starting at $p_{nff}(k)$, find the first nonfull frame. Update $p_{nff}(k) \leftarrow p_{nff}(k) + 1$, modulo N.

Figure 9.7 shows an example of the frames for output k. Here, the 3DQs are rearranged per EI order. The labels on top of the queues indicate the full or nonfull frame index, or the service order, and the numbers in the queue are the time stamps (e.g., arrival times). Initially, $p_{ff}(k) = p_{nff}(k) = 3$, and the frame pointers are as indicated in the figure. The first full frame (ff 1 as pointed by $p_{ff}(k)$) comprises cells with time stamps 136, 137, and 138. From here, p_{3k} points to II 1 (time stamp 136). The $p_{ff}(k)$ then moves to EI 1 (and so p_{3k} goes back to II 1), and after serving those cells, it moves to EI 2. Here, p_{2k} points to II 3 so the full frame consists of only 1 cell, the last of the frame. After no full frames are available, nonfull frames are served. After nff 1, nff 2, and nff 3 are served, $p_{ff}(k) = p_{nff}(k) = 3$, $p_{1k} = 2$, and $p_{2k} = p_{3k} = 3$.

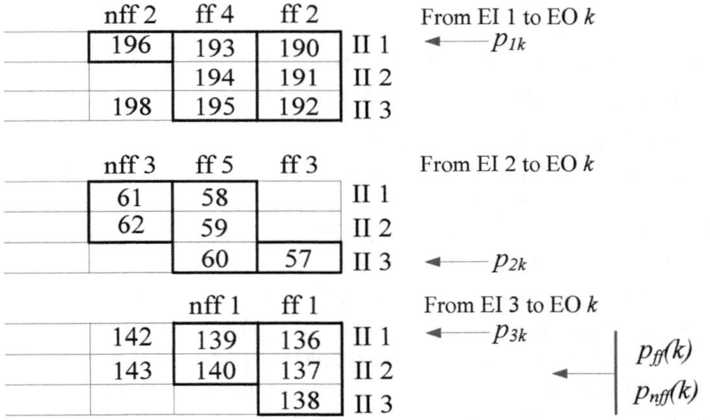

nff 2	ff 4	ff 2		From EI 1 to EO k
196	193	190	II 1	$\longleftarrow p_{1k}$
	194	191	II 2	
198	195	192	II 3	

nff 3	ff 5	ff 3		From EI 2 to EO k
61	58		II 1	
62	59		II 2	
	60	57	II 3	$\longleftarrow p_{2k}$

nff 1	ff 1		From EI 3 to EO k
142	139	136	II 1 $\longleftarrow p_{3k}$
143	140	137	II 2 $\longleftarrow$ $p_{ff}(k)$
		138	II 3 $\qquad p_{nff}(k)$

FIGURE 9.7
Example of FFF scheme for EO_k.

The two-stage switch using the FFF scheme achieves 100% throughput and keeps cells in order. However, there is some complexity associated with identifying full frames and frame sizes as they may need to be identified before frames are served.

9.5 Load-Balanced Combined-Input Crosspoint-Buffered Switches

The schemes for preventing or limiting the extent of out-of-sequence may need to coordinate among the different intermediate queues placed after the load-balancing stages. Buffered crossbars may be used to overcome such requirement as the fabric uses buffers inside.

In a combined-input crosspoint-buffered (CICB) switch, the feedback sent from a crosspoint to the input, indicating when a cell has left the crosspoint buffer, is critical to keep the system utilized. A common practice is to use larger buffer size to compensate for the time this communication takes. At the same time, it is more economical to keep buffer sizes close to a minimum. Therefore, sharing of crosspoint buffers by multiple inputs may be a desirable approach. Therefore, one of the advantages of this approach is the reduction on the amount of memory required. Two switches use this working scheme: the load-balanced CICB switch with full access to crosspoint buffers (LB-CICB-FA) and the load-balanced CICB switch with single access to crosspoint buffers (LB-CICB-SA) [142, 143]. These two switches are discussed in the following sections.

9.5.1   Load-Balanced CICB Switch with Full Access to Crosspoint Buffers (LB-CICB-FA)

The LB-CICB-FA switch has N VOQs in each input port, a fully interconnected stage (FIS) to interconnect one input to any of the N^2 crosspoint buffers, and a buffered crossbar. Figure 9.8 shows the architecture of the LB-CICB-FA switch. As before, the input ports are called external inputs, each denoted as EI_i. The outputs of the FIS are called internal outputs (IOs), each denoted as IO_l where $0 \leq l \leq N-1$, and they are also called internal inputs of the buffered crossbar (each is also denoted as II_l). The outputs of the buffered crossbar, or output ports, are called external outputs, each denoted as EO_j. A VOQ is again denoted as $VOQ(i,j)$. A crosspoint in the buffered crossbar is denoted as $CP(l,j)$ and the corresponding crosspoint buffer (CPB) is denoted as $CPB(i,j)$.

Different from the LB-CICB switch (Section 9.5), a crosspoint is not restricted to a one-to-one association with a VOQ and dedicated CPBs because the FIS enables input i to access any $CPB(l,j)$. The FIS can be implemented with an N-to-1 multiplexer, denoted as $MUX(l,j)$, per CPB, and the selection of an input and CPB can be resolved through a matching process, such that up to one cell can be written into a CPB in a time slot. We consider CPBs with a size of one cell, $k = 1$, and no memory speedup, in this section.

There are N^2 VOQ counters (VCs) at the buffered crossbar, denoted as $VC(i,j)$, one per $VOQ(i,j)$. Each VC keeps a count of the number of cells in the corresponding VOQ. There is an output access scheduler (OAS) per EO, denoted as OAS_j, and an input access scheduler (IAS) per EI, denoted as IAS_i, both placed in the buffered crossbar. IASs and OASs perform a parallel matching to determine which $CPB(l,j)$ receives a cell by selecting a row l for each j. The parallel matching is a distributed process where each IAS sends a request to all those OASs for which it has a cell, each OAS selects a request and grants the selected IAS, and each IAS selects a grant and sends an acceptance to the selected OAS. An IAS generates requests for its associated input based on the values of the VCs and accepts a grant for the input if

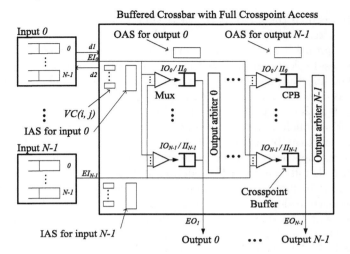

FIGURE 9.8
$N \times N$ load-balancing CICB switch with full access (LB-CICB-FA).

multiple grants are received. An OAS has a counter $RC(j)$ that counts the number of available CPBs for an output. An OAS grants IASs (inputs) that request access. A CPB is considered available if it has available space for one cell. The number of iterations to perform this match equals the minimum number of requests and $RC(j)$. After a matching process, $VC(i,j)$ and $RC(j)$ are updated. An output has an output arbiter that selects a CPB to forward a cell to the output among those occupied.

The LB-CICB-FA switch works as follows. A cell that arrives in input i and is destined to output j is stored in $VOQ(i,j)$. The input sends a notification of the cell arrival to the buffered crossbar, and the corresponding $VC(i,j)$ is increased by one after receiving this notification. In the next time slot, a request is sent from IAS_i to OAS_j. OAS_j selects up to N cells for crosspoints at output j after considering all requests from nonzero VCs and the availability of CPBs. The OAS grants the IAS whose requests are selected. Since an input may be granted access to multiple CPBs at different outputs (i.e., an IAS may receive several grants), the IAS accepts one grant and notifies the granting OASs. The IASs and OASs use either random or the longest queue first (LQF) as selection schemes in this section.

The LQF scheme is adopted to explore the maximum performance of the LB-CICB-FA switch under uniform and nonuniform traffic patterns under computer simulation, and random selection is adopted to analyze the maximum achievable throughput. The IAS and OAS may adopt other selection schemes in addition to random and LQF. This weighted selection is based on the VC values. After being notified by the buffered crossbar, an input sends the selected cell to the CPB at the next time slot. After a cell arrives at the CPB, the corresponding VC is decreased by one.

The output arbiter at output j (note that this is not part of the crosspoint-access process) selects an occupied crosspoint buffer to forward a cell to the output. The selection schemes can be random selection, round-robin, or FCFS. This switch does not require memory speedup.

Figure 9.9 shows an example of the operation of a 3×3 LB-CICB-FA switch. For simplicity, the FIS is represented with dashed lines between the input ports and the buffered crossbar, and it is assumed that arrived cells in the CPBs can be selected in the same time slot for dispatching. The selected paths that cells follow in the FIS from an input to CPBs are presented as solid lines. Figure 9.10 shows the matching process performed between the IAS and OAS at each time slot according to the VOQ occupancies in Figure 9.9.

At time slot t, as shown in Figure 9.9(a), there are six cells in the VOQs: A, B, C, D, E, and F. The VCs have the following values: $VC(0,0)=2$, $VC(0,2)=1$, $VC(1,0)=1$, $VC(2,1)=1$, and $VC(2,2)=1$. Because all CPBs are empty, $RC(0)=3$, $RC(1)=3$, and $RC(2)=3$. The matching process is performed between IASs and OASs as shown in Figure 9.10(a). In the request phase, IAS_0 sends requests to OAS_0 and OAS_2, IAS_1 sends a request to OAS_0, and IAS_2 sends requests to OAS_1 and OAS_2. In the grant phase, OAS_0 sends grants to both IAS_0 and IAS_1 as it receives two requests and $RC(0) = 3$, OAS_1 sends a grant to IAS_2, and OAS_2 sends grants to both IAS_0 and IAS_2. In the accept phase, IAS_0 sends an accept to OAS_0, IAS_1 sends an accept to OAS_0, and IAS_2 sends an accept to OAS_1. Cells A, C, and D are selected for forwarding in the next time slot. The corresponding VCs and RCs are updated to $VC(0,0)=1$, $VC(1,0)=0$, $VC(2,1)=0$, $RC(0)=1$, $RC(1)=2$, and $RC(2)=3$. The configuration of the interconnection stage is decided by the matching between IASs and OASs and on the selection of the CPBs. Here, available CPBs are selected randomly.

At time slot $t+1$, as Figure 9.9(b) shows, Cells A, C, and D are forwarded to $CPB(0,0)$, $CPB(1,0)$, and $CPB(2,1)$, respectively, where the FIS is configured to interconnect EI_0 to IO_0, EI_1 to IO_1, and EI_2 to IO_2. Matching for this time slot is performed as shown in Figure 9.10(b), and Cells B and E are selected. Because the only available CPB for Output 0 is $CPB(2,0)$, EI_0 is interconnected to IO_2 and EI_2 is interconnected to IO_1. Output arbiters perform FCFS scheduling to select cells to be forwarded to the output ports. Here, Cells A and D are selected to be forwarded to Outputs 0 and 1, respectively. These selections empty $CPB(0,0)$ and $CPB(2,1)$. The corresponding VCs and RCs are updated to $VC(0,0)=0$, $VC(2,2)=0$, $RC(0)=1$, $RC(1)=3$, and $RC(2)=2$.

At time slot $t+2$, as Figure 9.9(c) shows, Cells B and E are forwarded to $CPB(2,0)$ and $CPB(1,2)$, respectively, as the FIS interconnects EI_0 to IO_2 and EI_2 to IO_1. The matching for this time slot is performed as shown in Figure 9.10(c), where Cell F is selected (see Figure 9.9(c)). Cells A and D are forwarded to the output port. Output arbiters select Cells C and E for forwarding in the next time slot. The corresponding VCs and RCs are updated, $VC(0,2)=0$, $RC(0)=2$, $RC(1)=3$, and $RC(2)=2$.

At time slot $t + 3$, as shown in Figure 9.9(d), Cell F is forwarded to $CPB(0,2)$ as EI_0 is interconnected to IO_0. Cells C and E are forwarded to Outputs 0 and 2, respectively.

9.5.2 Load-Balanced CICB Switch with Single Access (LB-CICB-SA)

The LB-CICB-FA switch has the advantage of fully utilizing the CPBs of the buffered crossbar. However, the complexity of hardware implementation is high. The load-balanced CICB switch, abbreviated as LB-CICB-SA, was introduced to simplify the hardware while still allowing flexible access to CPBs.

The LB-CICB-SA switch has a simple load-balancing stage (LBS) that uses predetermined and cyclic configurations, VOQs at the inputs, and a buffered crossbar. Figure 9.11 shows the LB-CICB-SA switch. As in the LB-CICB-FA, the terms used are external input (EI_i) for an input port, internal output (IO_l) for an output of LBS, where $0 \leq l \leq N - 1$, internal input (II_l) for an input of the buffered crossbar, and external output (EO_j) for an output of the buffered crossbar. A crosspoint in the buffered crossbar that connects II_l to EO_j is denoted as $CP(l,j)$ and the crosspoint buffer as $CPB(l,j)$.

As in the LB-CICB-FA switch, there are N virtual counters, denoted as $VC(i,j)$, one for each input at the LBS in the LB-CICB-SA switch. In each EI, there is an input arbiter. In each II_l, there is one crosspoint-access scheduler (CAS), denoted as (CAS_l), that schedules the access to CPB_l (via II_l) for input i. Here, CPB_l represents the row of CPBs at II_l. A CAS and the input arbiter at EI_i selects a CPB and a VOQ, respectively, using LQF selection.

The LB-CICB-SA switch works as follows. At time t, the configuration of the LBS interconnects EI_i to II_l by using $l = (i+t)$ modulo N. At EI_i, a cell destined to output j arrives at $VOQ(i,j)$ and sends a notification to $VC(i,j)$ indicating the arrival. At the beginning of each time slot, each input arbiter sends a request to CAS_l as assigned by the configuration of the LBS. CAS_l selects a request from the nonempty VOQ with the longest occupancy for the available $CPB(l,j)$ and the inputs are notified.

The input dispatches the selected cell to the CPB in the next time slot. After that, the cell traverses the interconnecting stage and is stored at the CPB, and the corresponding VC is decremented by one. A cell going from EI_i to EO_j may enter the buffered crossbar through II_l and be stored in $CPB(l,j)$. Cells leave EO_j after being selected by the output arbiter. As in LB-CICB-FA, the output arbiters in LB-CICB-SA also use FCFS selection to arbitrate the forwarding of cells of flow $f(i,j)$ to output j.

An example of the operation of LB-CICB-SA. Figure 9.12 shows an example of the operation of a 3×3 LB-CICB-SA switch. The scheduling of cells takes place one time slot before the designated data path configuration is set up. At time slot t, the LBS is configured as shown in Figure 9.12(a). However,

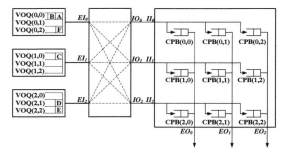

(a) Time slot t.

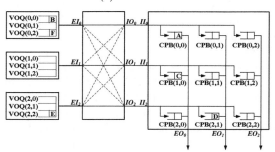

(b) Time slot $t + 1$.

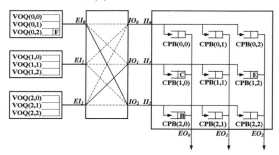

(c) Time slot $t + 2$.

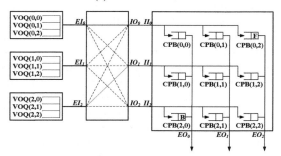

(d) Time slot $t + 3$.

FIGURE 9.9
Example of the operation of a 3×3 LB-CICB-FA switch.

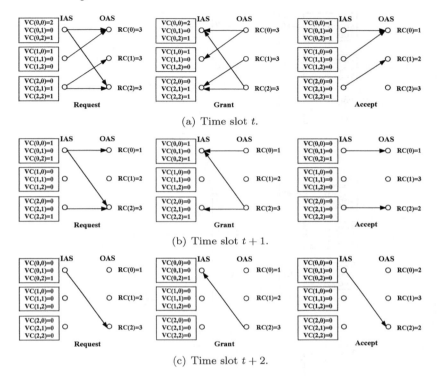

(a) Time slot t.

(b) Time slot $t + 1$.

(c) Time slot $t + 2$.

FIGURE 9.10
Matching process between IAS and OAS for the example of 3×3 LB-CICB-FA switch in Figure 9.9.

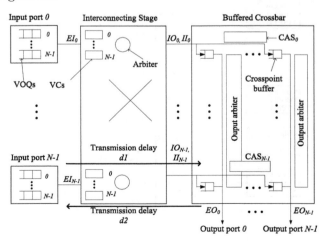

FIGURE 9.11
$N \times N$ load-balancing CICB switch with single access (LB-CICB-SA).

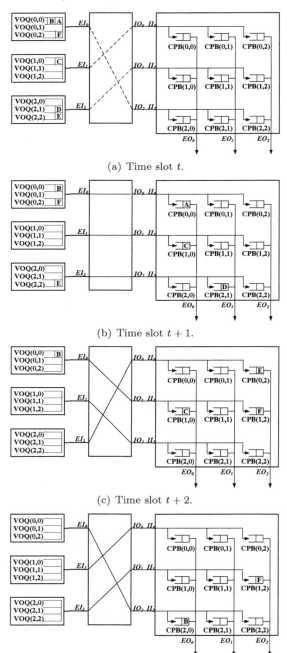

(a) Time slot t.

(b) Time slot $t + 1$.

(c) Time slot $t + 2$.

(d) Time slot $t + 3$.

FIGURE 9.12
Example of a 3×3 LB-CICB-SA switch.

since this is the first time slot, no cell is scheduled to use this configuration. The scheduling scheme considers the LBS configuration of the next time slot where EI_0 is interconnected to II_0, EI_1 is interconnected to II_1, and EI_2 is interconnected to II_2, as shown in Figure 9.12(b). Because all crosspoint buffers are empty, CAS_0 selects the head-of-line (HoL) cell of the longest queue, or Cell A. At the same time, CAS_1 selects Cell C (as there is no other cell in Input 1), and CAS_2 selects Cell D. This selection is arbitrary as the lengths of $VOQ(2,1)$ and $VOQ(2,2)$ are both equal to one cell.

At time slot $t + 1$, as shown in Figure 9.12(b), the selected Cells A, C, and D are forwarded to $CPB(0,0)$, $CPB(1,0)$, and $CPB(2,1)$, respectively. In this time slot, CAS_0 selects Cell E and CAS_1 (arbitrarily) selects Cell F to be forwarded in the next time slot. Output arbiters at Outputs 0 and 1 select Cells A and D to be forwarded to Outputs 0 and 1, respectively.

At time slot $t + 2$, as shown in Figure 9.12(c), Cells E and F are forwarded to $CPB(0,2)$ and $CPB(1,2)$, respectively. In this time slot, CAS_2 selects Cell B to be forwarded in the next time slot. Cells A and D are forwarded to the output ports as scheduled. Output arbiters at Outputs 0 and 2 select Cells C and E to be forwarded to Outputs 0 and 2, respectively, in the next time slot.

At time slot $t + 3$, as shown in Figure 9.12(d), Cell B is forwarded to $CPB(0,2)$. Cells C and E are forwarded to Outputs 0 and 2, respectively. Output arbiter at Output 2 selects Cell F to be forwarded in the next time slot.

9.5.3 Switch Performance Analysis

This section introduces a theoretical study of the stability of the LB-CICB switches. The analysis uses the LB-CICB-FA switch as an example to show the approach to analyze the stability of the switch performance. Similar to the theoretical analysis approaches in other chapters, the following conditions are considered in the analysis:

- The incoming traffic at the inputs is independent and identically distributed (i.i.d.).

- The arrivals at each input port and crosspoint buffer are Poisson processes.

- The selection of a nonempty VOQ at an input and the selection of a nonempty CPB per output are performed using a random selection.

- A CPB where an input forwards a cell of the selected VOQ is randomly selected. This is shown in Figure 9.13(a).

The performance of the LB-CICB-FA switch can be stated in the following theorem:

Theorem 1 *The LB-CICB-FA switch represented by the set of VOQs, where inputs under exogenous arrival processes can be assigned to a CPB, in the set*

$CPB(l,j) \ \forall \ 0 \leq l,j \leq N-1$, *randomly with uniform distribution among j, is weakly stable.*

Under a stationary exogenous arrival process A_n, a system of queues is weakly stable if, for every $\epsilon > 0$, there exists $B > 0$ such that $\lim_{n \to \infty} P\{||X_n|| > B\} < \epsilon$, where P_E denotes the probability of event E [58]. Weak stability implies rate stability where queue sizes are allowed to grow indefinitely with sublinear rate.

The selection of a VOQ, a CPB, and the configuration of the LBS follow a random selection policy. This policy is selected because of its analyzable properties, despite its expected modest performance. Here we use the following notations in the analysis:

- ρ_s – input load of the switch, $0 \leq \rho_s \leq 1$.

- $\lambda_{i,j}$ – average arrival rate of flow $f(i,j)$.

- $\lambda_{i,j}^X$ – average arrival rate at $CPB(i,j)$.

- $\mu_{i,j}$ – average service rate for $CPB(i,j)$.

- P_{si} – state probability that there are i cells in the queue.

- $P_{i,j}$ – transition probability from state i to state j, in other words, the transition probability from the state where there are i cells in the queue to the state that there are j cells in the queue.

- n – number of cells in a VOQ.

- k – CPB size.

A VOQ is modeled as an $M/M/1$ queue as the arrivals are Poisson processes and the service times received is exponentially distributed. Because arrivals are i.i.d., the VOQs of all N inputs for output j can be represented as a superposed Markov process with aggregated arrival rate

$$\lambda_j = \sum_{i=0}^{N-1} \lambda_{i,j} \tag{9.3}$$

This aggregated queue is represented as VOQ_j. VOQ_j can access all N CPBs for output j N times at each time slot. Therefore, it can be modeled as an $M/M/N$ queue, as Figure 9.13(b) shows. Here, $\rho = \frac{\lambda_j}{N\mu^I}$, where μ^I is the service rate of the aggregated M/M/N queue. The steady-state probability of n cells in VOQ_j is represented as P_{sn}, which is calculated using the following equations [13]:

$$P_{sn} = \begin{cases} P_{s0} \frac{(N\rho)^n}{n!}, & n \leq N \\ P_{s0} \frac{\rho^n N^N}{N!}, & n > N \end{cases} \tag{9.4}$$

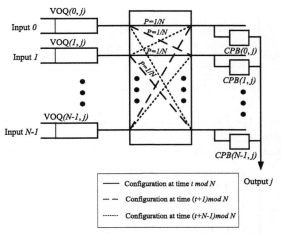

(a) Configuration of the load-balancing stage at different time slots.

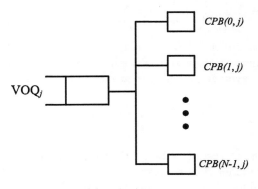

(b) M/M/N queue.

FIGURE 9.13
M/M/N queueing model of the LB-CICB-FA switch.

$$P_{s0} = \left[\sum_{n=0}^{N-1} \frac{(N\rho)^n}{n!} + \frac{(N\rho)^N}{N!(1-\rho)} \right]^{-1} \tag{9.5}$$

From (9.4) and (9.5),

$$P_{sn} = \begin{cases} \left[\sum_{n=0}^{N-1} \frac{(N\rho)^n}{n!} + \frac{(N\rho)^N}{N!(1-\rho)} \right]^{-1} \frac{(N\rho)^n}{n!}, & n \le N \\ \left[\sum_{n=0}^{N-1} \frac{(N\rho)^n}{n!} + \frac{(N\rho)^N}{N!(1-\rho)} \right]^{-1} \frac{\rho^n N^N}{N!}, & n > N \end{cases} \tag{9.6}$$

The service rate of the $M/M/N$ queue μ^I is determined by the availability of the CPBs for output j.

The state probabilities that there are n cells in a VOQ can be written as

$$P_{sn} = \begin{cases} \left[\prod_{t=1}^{n} \frac{\rho_s \cdot \lambda_j}{t \cdot \mu^I} \right] \cdot P_{s0}, & n \le N \\ \left[\frac{1}{N!} \cdot \left(\frac{\lambda_j}{\mu^I} \right)^N \right] \left[\prod_{t=N+1}^{n} \frac{\lambda_j}{N \cdot \mu^I} \right] \cdot P_{s0}, & n > N \end{cases} \tag{9.7}$$

$$P_{s0} = \left[\sum_{t=1}^{N-1} \frac{1}{t!} \cdot \left(\frac{\lambda_j}{\mu^I} \right)^t + \sum_{t=N}^{n} \frac{1}{N! N^{t-N}} \left(\frac{\lambda_j}{\mu^I} \right)^t \right]^{-1} \tag{9.8}$$

Each crosspoint buffer $CPB(i,j)$ is modeled as an $M/M/1/k$ queue. Because one of the motivations is to use small CPBs, the LB-CICB-FA switch is set with $k=1$. The average arrival rate at each CPB after the load-balancing stage is:

$$\lambda_{i,j}^X = \sum_{i=0}^{N-1} \lambda_{i,j} \frac{1}{N} \tag{9.9}$$

The calculation of the probability that $CPB(i,j)$ is available uses the $M/M/1$ queueing model. The superscript X in the following terms is used to represent the variables corresponding to the CPBs. The following probabilities are then defined:

P_{ij}^X – transition probability from state i to state j.

P_{si}^X – state probability that there are i cells in the CPB.

Here,

$$P_{01}^X = \sum_{i=0}^{N-1} \frac{1}{N} \cdot \frac{\lambda_{i,j}}{\sum_{j=0}^{N-1} \lambda_{i,j}} \cdot \rho \cdot \lambda_{i,j} \tag{9.10}$$

or

$$P_{01}^X = \frac{1}{N} \tag{9.11}$$

Because the output scheduler chooses a CPB with probability $\frac{1}{N}$,

$$P_{10}^X = \frac{1}{N} \tag{9.12}$$

From

$$\begin{cases} P_{01}^X P_{S0}^X = P_{10}^X P_{S1}^X; \\ P_{S0}^X + P_{S1}^X = 1; \end{cases} \tag{9.13}$$

$$P_{s0}^X = \frac{P_{10}^X}{P_{10}^X + P_{01}^X} \tag{9.14}$$

The arrival to each CPB after the load-balancing stage is $\lambda_{i,j}^X = \sum\limits_{i=0}^{N-1} \lambda_{i,j} \cdot \frac{1}{N}$, as stated before. Then, Equation 9.14 is $P_{S0}^X = \frac{1}{2}$.

The service rate of the aggregated queue can be approximated by the state probability when the CPB is available or there is no cell in the CPB, $\mu^I = P_{s0}^X$.

Under admissible i.i.d. traffic, $\sum\limits_{i=0}^{N-1} \lambda_{i,j} \leq 1$. Therefore, $\lambda_{i,j}^X \leq \frac{1}{N}$, $\lambda_j \leq 1$, $\rho_{max} = \frac{\lambda_j}{N\mu^I} = \frac{2}{N}$.

From Equation 9.7:

$$P_{sn} = \frac{N^{N-n}}{N! \left[\sum\limits_{t=0}^{N-1} \frac{2^{N-n}}{t!} + \frac{2^{N-n}N}{N!(N-2)} \right]} \tag{9.15}$$

As $n \to \infty$,

$$\lim_{n \to \infty} P_{sn} = 0 \tag{9.16}$$

The VOQ length n converges to ε, where $\varepsilon < \infty$, $\lim_{n \to \infty} P\{P_{sn} > B\} < \varepsilon$. Therefore, the weakly stable condition of the system of queues is met. It is proven that the LB-CICB-FA switch, with random selection, is weakly stable under admissible i.i.d. traffic.

9.6 Exercises

1. Why may a load-balanced Birkhoff–von Neumann switch suffer from forwarding packets in out of order?

2. Find the permutation matrices (switch configurations) to achieve 100% throughput in a 3×3 Birkhoff–von Neumann switch where the traffic has the following distribution profile:

$$\begin{bmatrix} 0.4 & 0.2 & 0.1 \\ 0.1 & 0.5 & 0.2 \\ 0.2 & 0 & 0.6 \end{bmatrix}$$

where the rows are the inputs and the column are outputs.

3. In the following 3×3 switch, the average traffic load between inputs and outputs can be presented as

$$\begin{bmatrix} 0.4 & 0 & 0.1 \\ 0.1 & 0.5 & 0.3 \\ 0.1 & 0.3 & 0.2 \end{bmatrix}$$

Show the permutations needed in a Birkhoff–von Neumann switch to provide 100% throughput to this traffic pattern.

4. Show the permutations needed for the following traffic matrix to be 100% served.

$$\begin{bmatrix} 0.4 & 0.2 & 0.1 & 0.2 \\ 0 & 0.3 & 0.2 & 0.1 \\ 0.2 & 0 & 0.2 & 0.3 \\ 0.4 & 0.1 & 0.3 & 0 \end{bmatrix}$$

5. Discuss what are the pros and cons on preventing out-of-sequence forwarding and limiting the number of cells forwarded in out of sequence.

6. Discuss how large should be the size of a CPB to avoid buffer underflow if the flow control mechanism is credit based.

7. Discuss what issue the flow splitter used in the FCFS scheme to limit the number of out-of-sequence cells solves.

10

Clos-Network Packet Switches

CONTENTS

The Clos network was proposed in 1952 to build large switches for circuit-oriented networks [42]. This network has been finding applications in communications networks ever since, including its application to packet switching. A Clos network is used to build very large switches using small switches.

The number of works on Clos networks is very large. This architecture has generated such a large number of designs and configuration schemes that the discussion of them would probably cover an entire book. This chapter describes some of the existing Clos-network packet switches and configuration schemes.

10.1 Introduction

Clos networks find their application in the design and construction of large switches. These networks were originally designed for circuit switching but they are also applicable to packet switching. The Clos principle uses small switches, called switching modules (modules for short), to build larger switches. Clos networks potentially reduce the amount of hardware needed to build a large switch. A Clos network provides multiple paths to interconnect an input to an output and they may be blocking. The number of path choices combined with its blocking property can make a Clos-network switch complex to configure, and unsuitable configurations may affect the switch's performance.

10.2   Clos Networks

A Clos network may have three stages of interconnected modules. Each module is a switch, such as a crossbar. The first stage of the switch comprises k modules. Each module has n inputs, such that $kn = N$, where N is the number of input/output ports of the Clos-network switch, as Figure 10.1 shows. Each module in a stage has one link interconnecting it to a module in the adjacent stage. That is, modules in the first stage interconnect with each module in the second stage, every module in the third stage interconnects with each module in the second stage, and each module in the second stage interconnects with each module in the first stage and each module in the third stage. The modules in the first stage have m outputs to connect to m modules in the second stage. The value of m can be selected arbitrarily, but a small number makes the switch blocking, as discussed below. A module in the second stage has k inputs and as many outputs as the number of modules in the third stage. If the switch also has N outputs, and each module in the third stage has n outputs, then the number of outputs of second-stage modules is k. A third-stage module has m inputs.

First-, second-, and third-stage modules are also called input modules (IMs), central modules (CMs), and output modules (OMs). A Clos network then has k $n \times m$ IMs and OMs, and m $k \times k$ CMs in an $N \times N$ switch. The description of the different switches presented in the remainder of this chapter uses these variables as well. Also, these packet switches segment variable-size packets into fixed-size packets, referred to as cells, when the packets arrive in the switch. Packets are re-assembled before they leave the switch.

Therefore, n defines the number of IMs, OMs, and the size of CMs. The number of CMs can be decided arbitrarily. Figure 10.2 shows examples of Clos-network switches with different numbers of CMs. For example, Figure 10.2(a) shows a 6×6 switch with a single CM. IM and OM modules are interconnected through the CM but if all the inputs of an IM have packets for all three outputs of an OM, blocking would occur. Therefore, a small number of CMs can make the switch blocking. Figure 10.2(b) shows a switch with three CMs and the blocking example in the switch with a single CM (Figure 10.2(a)) may not occur here. Figure 10.2(c) shows a 6×6 switch with $n = 2$ so

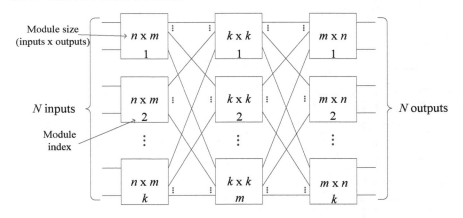

FIGURE 10.1
Three-stage Clos-network switch.

that CMs are 3×3 switches. Figure 10.2(d) shows an example of a switch with different numbers of inputs and outputs. In this case, this is a 6×9 switch. The size of the OMs, 3×3, is different from the size of IMs, 2×3. Yet, the switch has the same number of IMs and OMs. The remainder of this chapter considers $N \times N$ switches, unless otherwise stated.

Clos-network switches may be blocking because the number of CMs (paths to the outputs) may not be large enough or because, under limited paths, some routing decisions are not efficient, leading to overuse of some paths and underuse of others.

An advantage in using a Clos network for building large-size switches is that the number of switching elements in the switch is smaller than that in a crossbar (or bi-dimensional array of crossbars) switch of equivalent size. These savings in hardware are also reflected in the number of switch modules. This advantage seems to suggest that it is economical to consider a small number of CMs in a Clos-network switch. However, as discussed before, a small number of CMs may be prone to blocking. In fact, the example in Figure 10.3 shows $m = n$ and yet the switch is blocking.

This raises the question: How many CMs are needed in a Clos-network switch to make it nonblocking? A sufficient condition to make a Clos network nonblocking is the following. Consider the switch in Figure 10.4. This figure shows the connections between a single IM and single OM. The IM has n inputs and the OM has n outputs. Let us assume that the IM and OM have only one idle port each, to be interconnected. This means that $n - 1$ of the output ports of the OM are busy, so that there are $n - 1$ CMs with their links interconnected to the OM that are occupied. In a similar way, the IM has $n-1$ output links, interconnected to $n - 1$ CMs, occupied. Therefore, an additional

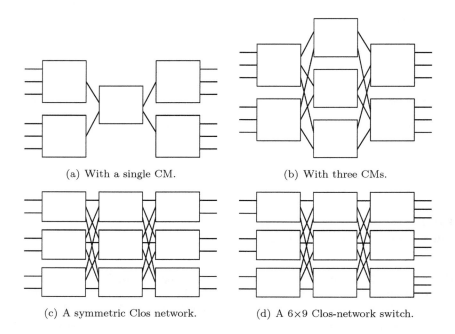

(a) With a single CM.   (b) With three CMs.

(c) A symmetric Clos network.   (d) A 6×9 Clos-network switch.

FIGURE 10.2
Examples of Clos-network switches with different numbers of CMs and ports.

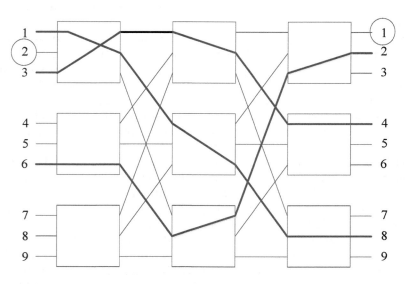

FIGURE 10.3
Example of blocking in a 9 × 9 Clos-network switch.

CM, in addition to those busy ones, is needed. That is,

$$m \geq 2(n-1) + 1 = 2n - 1 \tag{10.1}$$

Therefore, such value of m is a sufficient condition to have a nonblocking switch.

In such a case, the number of crosspoints in a Clos-network switch, N_x, where the modules are crossbar switches, is the sum of knm crosspoints in the first stage, mk^2 crosspoints in the second stage, and knm crosspoints in the third stage, or

$$N_x = 2Nm + m\left(\frac{N}{n}\right)^2 \tag{10.2}$$

For a comparison, an $N \times N$ crossbar switch has N^2 crosspoints. For example, for $N = 1024$ ports and $n = 32$, $m \geq 63$. Let's consider $m = 64$. In this case, $N_x = 2^{17} + 2^8$ while $N^2 = 2^{20}$. Therefore, the number of crosspoints in a Clos-network switch is smaller than that in a crossbar switch for an equivalent N (assuming that a crossbar of a large size may be feasible to build).

10.2.1 Clos Network with More than Three Stages

The Clos-network concept can be recursively applied to large modules. This approach is useful when switch modules cannot be built of a large size. In case a module is implemented as a Clos network, the number of stages in the switch may be increased by two. Moreover, for a very-large-size switch and small-size modules, k may become large, and so the required size of CMs. Therefore, the CMs are the modules that most likely need to be built as Clos networks under such a scenario.

If that's the case, a three-stage switch then becomes a five-stage switch, and so on. Figure 10.5 shows an example of a five-stage Clos-network switch, where the Clos-network concept is re-applied to the CMs.

It is clear then that Clos-network switches may have three, five, seven, and so on, stages. However, it may be practical to consider up to five stages to build large switches. Some of the complexity that comes with a large number of stages is the configuration of the switch, the real-state required, the number of cables, and the time to configure the switch, to mention some of them.

10.3 Queueing in Clos-Network Switches

As any other packet switches, Clos-network switches are subject to output contention. Therefore, queueing may be needed to keep those cells (as segments of variable size packets) that lose contention waiting until they can

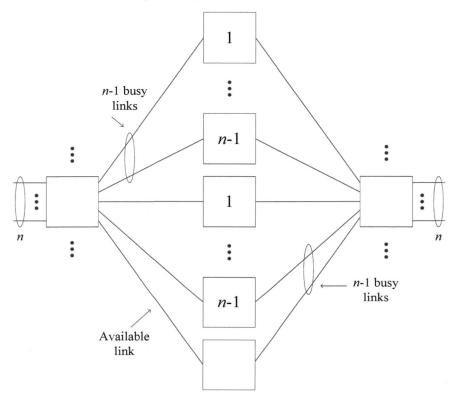

FIGURE 10.4
Condition to make a Clos network strictly nonblocking.

be forwarded to their destined outputs. Clos-network switches may have input or output queues, as in single-stage switches. They also can be internally buffered. The internal buffers may be placed in the modules of any of the different stages, as Figure 10.6 shows.

The name of the switch can be labeled to indicate the stage that has buffers; for this, modules that have no memory as denominated as switches that perform space switching (Space) and those with buffers as switches that perform queueing (Memory), in addition to switching. All the Clos networks in Figure 10.6 have three stages. Figure 10.6(a) shows a network with buffers in IMs and OMs, called a memory-space-memory (MSM) switch, for short.

The allocation of buffers in different stages affect the scheme and timing used to configure the complete switch (modules in the three stages). The timing may be affected by where in the switch configuration decisions take place. For instance, a Space-Space-Space (or S^3) switch must have all modules configured before a cell passes through. This process needs to perform several

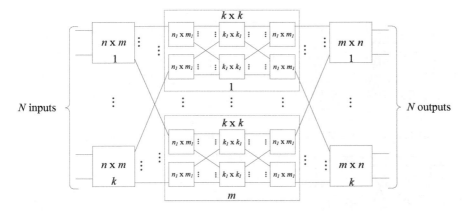

FIGURE 10.5
Clos-network switch with five stages.

matchings (between inputs and outputs of a module) and needs a long time for deciding and performing the configuration before a cell is transmitted through.

10.3.1   Space-Space-Space (S^3) Switches

S^3 switches are a basic architecture of a Clos network for packet switching as modules have no buffers. This switch requires that all modules be configured before any cell is forwarded through it. There are several schemes proposed for configuration of S^3 switches, usually referred to as scheduling or matching schemes. Because there are multiple paths to interconnect an input port to an output port, the central link (or module) has to be selected in addition to matching the ports. The design of a scheduling scheme must be done carefully to avoid blocking in switches. This requirement becomes stricter for a switch with a small number of CMs. Note that the condition in Inequality 10.1 states sufficiency. Therefore, it is possible that a Clos-network switch with a smaller m and an effective scheduling scheme would make the switch nonblocking.

In general, some schemes perform port matching first and the selection of the input-output path (or CM) follows. As described below, there is also a scheme that selects the CM first, using information about the ports, and then proceeds with port matching.

10.3.1.1   Port-First Matching Scheme

There are several schemes to configure S^3 switches that perform port matching first. Examples of those are the f-MAC and c-MAC schemes [32]. These schemes use frames as a scheduling time unit and as the method for cell transmission. A frame here is a group of cells in the same virtual output

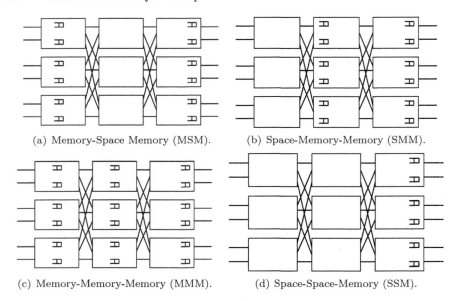

(a) Memory-Space Memory (MSM).   (b) Space-Memory-Memory (SMM).

(c) Memory-Memory-Memory (MMM).   (d) Space-Space-Memory (SSM).

FIGURE 10.6
Some queueing strategies in Clos-network switches.

queue (VOQ). The matching schemes used are based on schemes developed for single-stage input-buffered switches [105].

10.3.1.2   Module-First Matching (MoM) Scheme

An early method that uses module-first matching is path switching [101]. Here, traffic from input ports are aggregated to determine the traffic from IMs and OMs. A selection of CMs to carry those IM-OM matching follows. The scheme presents the selection of CMs as a color assignment problem. A recent scheme that uses module-first matching is the weighted module first matching (WMoM) scheme, which follows.

Weighted Module First Matching (WMoM) scheme. WMoM is proposed to reduce the complexity of building a scheduler for large S^3 Clos-network switches. WMoM matches modules first and then it matches ports from matched modules. WMoM adopts a weighted selection policy to achieve high switching performance under nonuniform traffic. This scheme uses two schedulers, one being a module-matching scheduler, S_M, that determines the module pairs. The other is a port-matching scheduler, S_P, that determines the port pairs. The weighted selection policy is the longest queue-occupancy first, which is similar to the iLQF algorithm [113]. However, other weight-based selection schemes (e.g., oldest cell first) can also be used. WMoM considers

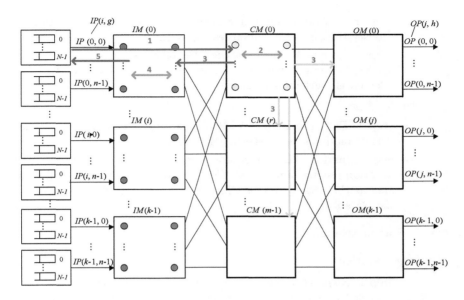

FIGURE 10.7
Clos-network switch with five stages.

the occupancy of an IM for module matching, as weight. Figure 10.7 shows the S^3 switch and the notations for different parts of the switch.

To perform module matching, WMoM uses a VOQ module counter (VMC) to count the number of cells in an IM for each OM. Each matching process follows a request-grant-accept approach [8]. In this S^3 switch, input ports have N VOQs, where N is the number of ports of the switch. Assuming a distributed implementation of the schedulers, one of the CMs host the S_M, and each IM has a S_P, as S_M performs global matching (i.e., among IMs and OMs) and S_P performs matching in an IM. Therefore, S_M has $2k$ arbiters, one per each CM input port and one per each CM output port, each called input-module arbiter and output-module arbiter, respectively. Each S_P has $2n$ arbiters, one for each IM input port and one per each OM output port, which are called input- and output-port arbiters, respectively.

The following is the description of WMoM:

- Part 1: Module matching

  - First iteration

    * Step 1. *Request*: Each VMC, whose count is larger than zero, sends a request to each output-module arbiter for which the requesting IM has at least a cell. A request carries the number of cells for that OM as weight.

> ∗ Step 2. *Grant*: If an unmatched OM receives any requests, it chooses the one with the largest occupancy.
>
> ∗ Step 3. *Accept*: If an unmatched IM receives one or more grants, it accepts the one with the largest occupancy. The matched CM selected is marked as reserved.

- *i*th iteration

  > ∗ Step 1. Each unmatched VMC sends a request to all unmatched output modules, as in the first iteration.
  >
  > ∗ Steps 2 and 3. The same procedure is performed as in the first iteration among unmatched VMCs and output-module arbiters as long as the corresponding CM has paths available to interconnect the IM and OM.

- Part 2: Port matching

  - First iteration

    > ∗ Step 1. *Request*: Each nonempty VOQ of the matched IM sends a request to each output port of the matched OM for which it has a queued cell, indicating the number of cells in that VOQ.
    >
    > ∗ Steps 2 and 3. *Grant* and *Accept*: The same procedure as in the module matching is performed for matching nonempty VOQs of a matched IM and OPs of a matched OM, by input port arbiters and output port arbiters at the IM. These arbiters select the requests with the largest occupancy. Ties are broken arbitrarily.

  - *i*th iteration

    > ∗ Step 1. Each unmatched VOQ in an IM at the previous iterations sends another request to all unmatched OP of the matched OM as in Step 1 of the first iteration.
    >
    > ∗ Steps 2 and 3. The same procedure is performed as in the first iteration for matching between unmatched nonempty VOQs and unmatched output ports in the matched IMs and OMs.

As the description of the scheme indicates. The scheme has two iteration levels, one for module and port matching, or local iterations, and the other for executing the whole scheme again, or global iterations.

WMoM achieves 100% throughput under uniform Bernoulli traffic with several global and local iterations, and the throughput under unbalanced (nonuniform) traffic [139] approaches 100%. The time needed for performing the matching process and executing the number of iterations needed defines the time needed to configure the switch.

10.3.2 Memory-Space-Memory (MSM) Switches

MSM switches have been considered in the past as a more feasible implementation of large switches [36]. An $N \times N$ MSM switch has buffers at the IMs and OMs, and some of the proposed architectures use N buffers (or queues) in each module. The queues in the IM are VOQs, where each one stores cells destined to a specific output port. The OMs have output queues, one per output. The switch has $n = m = k$, for simplicity. Note that the switch is not strictly nonblocking in this case. Figure 10.8 shows an MSM switch. In this switch, cells incoming to the IM are stored in the IM queues (VOQs). An IM queue is matched to a CM (or outputs of the IM) using random selection. So each IM queue can receive up to n cells and forward one cell.

The configuration of the CM follows a matching between the requests received from IMs to the destined OMs. The destination of a cell is used to determine which CM output is requested for matching. Once the CM configuration is decided, the paths from IM to OM is set and the cell from the selected IM queue is forwarded.

The need for matchings at IMs and CMs requires an arbiter for each matching party. A distributed implementation of the scheme would need arbiters at IM queues and outputs of the IMs, and arbiters for inputs and outputs of the CMs. As the matching at CMs need to be performed after the matchings at IMs are completed, the time to set the configuration of the switch is the time it takes to perform matching at IM, the propagation time the request takes to travel from an IM to a CM, the matching time at CM, and the propagation time to IM and inputs, to inform which matchings were successful (grants). The largest time of the configuration time and the cell transmission time sets the minimum duration of a time slot.

However, random selection may not be as effective as some requests from the IM queues may converge to the same IM outputs. For example, consider the small switch in Figure 10.9. In this 4×4 switch, all selections are performed randomly with a uniform distribution. Here, let's assume that queues have large backlogs so that a queue participates in matchings every time slot. To estimate the throughput of the switch, we may calculate the contribution of a single output port, Output port 0 ($OP(0,0)$ is the port in $OM(j = 0)$ and port $h = 0$ as $OP(j, h)$), by calculating the traffic it can dispatch. This calculation considers the queues for that output.

$IM(0)$ and $IM(1)$ have VOQs that may want to send traffic to OM(0), where $OP(0,0)$ is Output port 0. So, we can also consider the traffic passing through the link from $CM(0)$ to $OM(0)$. Queue $VOQ(i, j, h)$ stores cells at $IM(i)$ for $OP(j, h)$. Here, $VOQ(0, 0, 0)$ is a queue for $OP(0, 0)$, and it contends with another VOQ in $IM(0)$, $VOQ(0, 0, 1)$, for using the link from $IM(0)$ to $CM(0)$, and with two VOQs for OM in $IM(1)$, $VOQ(1, 0, 0)$ and $VOQ(1, 0, 1)$ for use of the link from $CM(0)$ to $OM(0)$. The amount of traffic

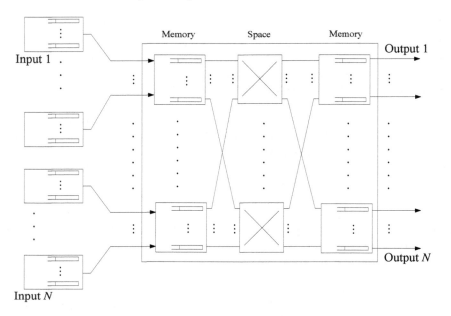

FIGURE 10.8
Example of an MSM switch.

that $VOQ(0, 0, 0)$ can send to Output port 0 is

$$\left(\frac{1}{4}\right)\left(\frac{1}{2}\right)\left(\frac{1}{2}\right) + \left(\frac{1}{4}\right)\left(1 - \frac{1}{2}\right)(1.0) = \frac{3}{16}$$

and considering this to be the same contribution for one of two CMs and one of two VOQs for that output port, the total throughput is

$$\left(\frac{3}{16}\right)(2)(2) = \frac{3}{4}$$

In general, the maximum achievable throughput, Thr, of such a switch may be calculated as [128]:

$$Thr = min\{\frac{m}{n}\sum_{i=0}^{k-1}\binom{k-1}{k-i-1}\frac{1}{k}^{k-i-1}\left(1 - \frac{1}{k}\right)^i\frac{1}{k-i}, 1.0\} \qquad (10.3)$$

10.3.2.1   Round-Robin Matching in MSM Switches

As Equation 10.3 shows, there is blocking in an MSM switch using random selection, and throughput degradation occurs as a result. To compensate that performance degradation a expansion on the middle stage of the switch would

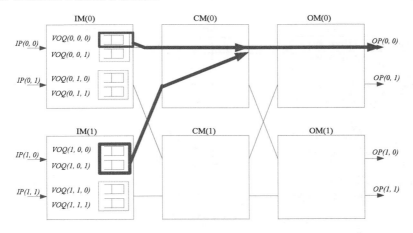

FIGURE 10.9
Example of throughput in a 4×4 MSM Clos-network switch using random
selection schemes.

be required. However, this expansion would also increase the cost of the switch.
To increase the throughput of a MSM switch without extension (i.e., $k = n = m$), round-robin was proposed as selection policy in the concurrent round-robin dispatching (CRRD) scheme [127].

Here, $L_i(i, r)$ is the output link of $IM(i)$ that is connected to $CM(r)$, and $L_c(r, j)$ is the output link of $CM(r)$ that is connected to $OM(j)$. Also, $L_i(i, r)$ and $L_c(r, j)$ have arbiters and pointers $P_L(i, r)$ and $P_c(r, j)$, as the arbiter uses round-robin as selection policy.

The operation of CRRD includes two phases. In Phase 1, CRRD employs an iterative matching to match VOQs and the output links of IMs (interconnected to CMs). A request associated with the matched output link is sent from the IM to the CM. In Phase 2, a selection of request for the output link of CMs (to OMs) is performed and the CMs send the arbitration results to the IMs.

- Phase 1: Matching within IM

    - First iteration

        Step 1. Each nonempty VOQ sends a request to every output-link arbiter.

        Step 2. Each output-link arbiter chooses one request in a round-robin fashion by searching from the position of $P_L(i, r)$. It then sends the grant to the selected VOQ.

        Step 3. The VOQ arbiter accepts a grant among all those received in a round-robin fashion by searching from the position of the VOQ pointer, $P_V(i, v)$, and informs the granting output-link arbiters.

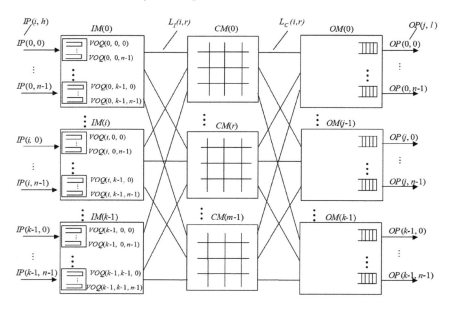

FIGURE 10.10
MSM Switch with round-robin-based dispatching scheme (CRRD).

> — ith iteration
>
>> Step 1. Each unmatched VOQ at the previous iterations sends another request to all unmatched output-link arbiters.
>>
>> Steps 2 and 3. The same procedure is performed as in the first iteration for matching between unmatched nonempty VOQs and unmatched output links.

- Phase 2: Matching between IM and CM

  Step 1. $L_I(i, r)$ sends a request of the matched VOQ-L_I to $CM(r)$. A round-robin arbiter for $OM(j)$ selects the request closer to $P_c(r, j)$ sorted in a round-robin schedule, and sends the grant to $L_I(i, r)$.

  Step 2. The IM that receives a grant from the CM forwards a cell from the corresponding VOQ the next time slot.

The CRRD scheme, as it is based on round-robin selection with pointer updates, as in iSLIP [111], achieves 100% throughput under uniform traffic without requiring expansion of the central stage of the Clos network. However, the VOQs in IMs and the output queues in OM need speedup, and speedup in memory is costly.

10.3.3 Memory Memory Memory (MMM) Switches

The MMM switch is an architecture that may require the shortest time for configuring a Clos-network switch. This switch allows to perform the selection of a queue that will forward a cell without waiting for other scheduling decisions from other modules, in other stages. Therefore, the time slot used by an MMM switch is short. Furthermore, the combined input-crosspoint buffered (CICB) switch [139] showed that crosspoint buffers may improve the switching performance of a switch, and an MMM switch is expected to achieve a high switching performance.

However, the use of memory in all stages of a Clos network, and in particular in the middle stage, creates some challenges. Because a Clos-network switch provides multiple paths to forward a cell from an input to an output, the use of buffers in the middle stage of an MMM switch can potentially cause forwarding of cells out of sequence, as different input-output paths may experience different queueing delays. Out-of-sequence forwarding requires resequencing of cells (and IP packets) at the outputs, and this process can be costly in terms of memory, packet processing time, and cause unnecessary retransmissions after triggering the fast-recovery mechanism of the Transmission Control Protocol (TCP) [54].

Figure 10.11 shows an example of an MMM switch and the queues throughout the switch. The queues in a module are similar to those of a buffered crossbar, where there is one queue for each input and output. These queues store cells for the module or ports of the following stage. For instance, input ports have VOQs, one per output of the switch. An IM has virtual central module queues (VCMQs), which store cells for different CMs, and from different inputs (CM queue per input). A CM has virtual output module queues (VOMQs), which store cells going to different OMs. An OM has virtual output port queues (VOPQs), which store cells for each port. Also, as the figure shows, there is an arbiter for each link to decide which queue forwards a cell to the next stage.

Another issue with this switch is that head-of-line (HoL) blocking may occur in the middle stage of the switch. Figure 10.12 shows two examples of HoL blocking in a 4×4 MMM switch. One of these cases occurs may occur at CMs (Figure 10.12(a)) for cells from different inputs and destined to different outputs, and the other case may occur at IMs (Figure 10.12(b)) for cells from the same input but destined to different outputs.

In the HoL blocking at CM, Figure 10.12(a) shows Cells A and B, from $IP_{2,1}$ and $IP_{2,2}$, respectively, destined to $OP_{2,1}$ and $OP_{2,2}$, respectively. They are both scheduled to go through $CM(2)$. Cell A in $VCMQ(2,1,2)$ is first scheduled to be forwarded to CM_2 and it is then stored at $VOMQ(2,2,2)$. Cell B, scheduled after Cell A, is placed behind Cell A in $VOMQ_{2,2,2}$. Cell A cannot be forwarded to OM_2 because its destined $VOPQ_{2,2,1}$ has no room available. Cell B could be forwarded to $VOPQ(2,2,2)$ as it has room available but it cannot be forwarded because cells A blocks it.

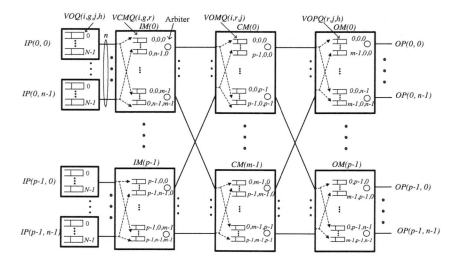

FIGURE 10.11
MMM switch with buffered crossbars as modules.

In the HoL blocking at IM, Figure 10.12(b) shows Cells C and D, both of them sent from $IP_{2,1}$ and destined to $OP_{1,1}$ and $OP_{2,2}$, respectively. They are both scheduled to go through CM_2. Cell C in $VCMQ_{2,1,2}$ must then be forwarded at $VOMQ_{2,2,2}$ but the queue has no room available. The destined queue for Cell D, $VOMQ_{2,2,1}$, however, is available. Yet, Cell C blocks Cell D. As these examples show, an MMM switch may suffer from HoL blocking, and this could affect the switch's throughput.

10.3.3.1   MMM with Extended Queues (MM^{e}M)

The VOQ principle is applied to avoid the HoL blocking phenomenon of an MMM switch, in what is called the MMM switch with extended memory, or MM^{e}M switch [54, 55]. Figure 10.13 shows the architecture of this switch, where the CMs have a new design. The CMs have per-output flow queues. In fact, per-input -output flow queueing would be the ideal case, but that would make the number of queues too large.

An MM^{e}M switch, using round-robin as arbitration scheme, achieves 100% throughput under uniform traffic, and approaches 100% under unbalanced nonuniform traffic for the minimum queue size (room for one cell). The performance of this switch improves as the switch size increases [54].

MM^{e}M does not consider forwarding of out-of-sequence cells, and it may still occur in this switch. Avoidance of out-of-sequence forwarding has been of interest. In a recent approach, a hash function was used to select the path that cells of flow would follow in the switch. However, speedup has been reported as required [31]. Other works consider synchronization of all inputs and each

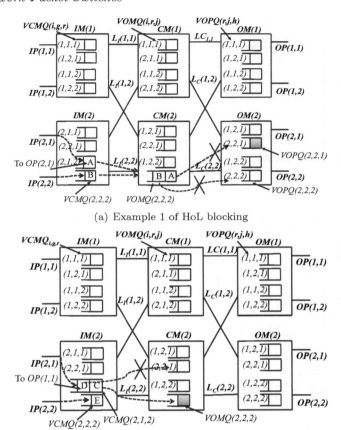

(a) Example 1 of HoL blocking

(b) Example 2 of HoL blocking

FIGURE 10.12
Examples of HoL blocking in a CM and IM of MMM switch.

of the stages of the switch, but it may be hard to accomplish as a large-size switch may need to place boards and parts apart, where the propagation delays may be varied and nonnegligible.

10.3.4 Space-Space-Memory (SSM) Switches

The SSM switch uses buffers in the modules at the third stage [15]. Figure 10.14 shows an example of this switch. The SSM switch has no buffers in the middle-stage modules, therefore it reduces the possibility of forwarding packets out-of-sequence. In addition, the use of buffers in the last-stage modules, the OMs, rids the configuration scheme of performing port matching, which is complex in a switch with a large number of ports, as in a Clos-network

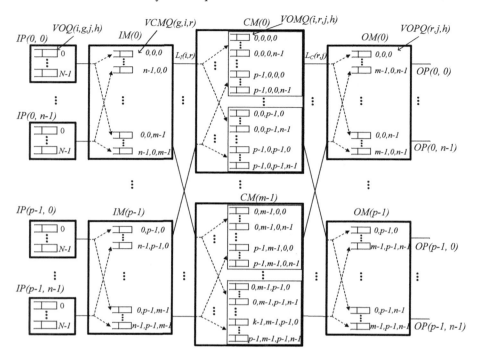

FIGURE 10.13
MMM switch with extended memory (MM^{e}M).

switch. However, it is important to use an efficient mechanism to configure IMs and CMs. For that, an SSM switch may use the weighted central module link matching (WCMM) scheme [15, 148].

10.3.4.1   Weighted Central Module Link Matching (WCMM)

The weighted central module matching (WCMM) scheme matches inputs to the output links of IMs. It then proceeds with matching input ports to the output links of CMs. Note that matching between input and output ports is not needed as the OM has buffers (i.e., crosspoint buffers). The CM-link matching process solves both matching IMs to OMs and selecting a CM to route, at the same time. Figure 10.14 shows the SSM switch and the arbiters used in WCMM. This example shows the CM-link arbiters in a CM and the IM-link arbiters in IMs. The input port and the input link of an IM are the same link, and it is denoted as the input port $IP(i, g)$. The output link of an IM, which is also the input link of the connecting CM, is denoted as $L_I(i, r)$. Similarly, a CM output link is denoted as $L_C(r, j)$. The arbiters in an IM are IM input link arbiter or input arbiter at $IP(g, i)$, and IM output link arbiter or

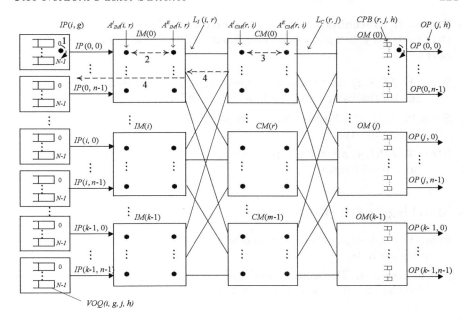

FIGURE 10.14
SSM Clos-network switch.

arbiter at $L_I(i, r)$ (for brevity the arbiters are denoted similarly). The arbiters in the CM are CM input-link arbiter $A_{CM}^I(r, i)$ and CM output-link arbiter $A_{CM}^E(r, i)$.

WCMM uses two matching phases. In Phase 1, WCMM performs iterative matching between $IP(g, i)$ arbiters and IM output-link arbiters. This process takes place at IMs. WCMM employs the longest queue first (LQF) selection policy in this matching process.

In Phase 2, WCMM performs an iterative matching between $A_{CM}^I(r, i)$ and $L_I(i, r)$ arbiters. The selection policy in WCMM is also based on LQF. The weights of $L_I(i, r)$ and $L_C(r, j)$ are determined by a VOQ module counter (VOMC), denoted as $VOMC(i, g, j)$, where

$$|VOMC(i, g, j)| = \sum_{h=1}^{n} |VOQ(i, g, j, h)|$$

This counter indicates the number of cells in $IP(i, g)$ that are destined to $OM(j)$. Matching processes also follow a request-grant-accept approach [8]. WCMM is described as follows:

Input arbitration. An input selects the VOMC with the largest weight and sends the information to the corresponding IM. Ties are broken arbitrarily.

- **Phase 1: Matching in IM: first iteration.** This phase defines the CM that routes a cell from a matching input to the destined OM (and output port).

    Step 1. $IP(i,g)$ arbiter sends a request to all $L_I(i,r)$ arbiters. Each request carries the value of VOMCs as weight value.

    Step 2. $L_I(i,r)$ arbiter selects the request with the largest weight. Ties are broken arbitrarily.

    Step 3. $IP(i,g)$ arbiter accepts the grant with the largest weight. Ties are broken arbitrarily.

- **Matching in IM: qth iteration**

    Step 1. Each unmatched input sends another request to all unmatched IM link arbiters.

    Steps 2 and 3. The same procedures are performed as in these two steps for the first iteration among unmatched links.

- **Phase 2: Matching CM links: first iteration**

    Step 1. After Phase 1 is completed, each matched $L_I(i,r)$ arbiter sends a request to its destined CM output-link arbiter, $A_{CM}^E(r,j)$. Requests include the number of cells for output link $L_C(r,j)$.

    Step 2. CM output-link arbiter $A_{CM}^E(r,j)$ selects the request with the largest occupancy and sends the grant to the corresponding CM input-link arbiter. Ties are broken arbitrarily.

    Step 3. CM input-link arbiter $A_{CM}^I(r,i)$ accepts the grant with the largest occupancy. Ties are broken arbitrarily.

- **Matching CM links: qth iteration**

    Step 1. Each unmatched CM input-link arbiter sends a request to all unmatched CM output-link arbiters.

    Steps 2 and 3. Matching of unmatched CM link arbiters is performed as in the first iteration.

Selection of VOQs and crosspoint buffers. After Phases 1 and 2 are completed, each input selects the VOQ with the largest occupancy from the selected VOMC group $(OM(j))$ to forward a cell to the destined crosspoint buffers (CPBs), each denoted as $CPB(r,j,h)$. The output arbiter at $OP(j)$ selects a cell among nonempty CPBs to forward a cell to the output port using the oldest-cell first (OCF) policy. Ties are broken arbitrarily.

Figure 10.14 shows the SSM switch with WCMM scheduling. The matching stages are numerated. In Step 1, each input selects the VOQ with the largest weight. In Step 2, an input is matched to an IM output link. In Step 3, each IM output link arbiter issues a match request for a given OM (according to the selected VOQ at IP) to CM arbiters. In Step 4, the arbiter at an OP selects the CPB with the oldest cell for forwarding to the output port.

Figure 10.15(a) shows how L_I links can be matched to different OMs. Here, L_I links are grouped by the CM they are connected to. As the figure shows, group of links A, belonging to different IMs, are only matched to CM links of the same CM, and these CMs can be connected to any OM. The complexity of matching different inputs to different outputs is simplified by matching the different link groups independently and simultaneously. The small black circles in Figure 10.15(b) represent IM output link and CM output links arbiters, which perform matching of these links at a CM.

Figure 10.16 shows an example of WCMM's Phase 1 matching in a 4x4 switch with $n = m = k = 2$. The figure shows the VOMC values at the inputs. In the request step, each input selects the VOMC with the largest value and sends the request to both L_I arbiters. At $IM(0)$, $IP(0,0)$, and $IP(0,1)$ select $VOMC(0,0,0)$ and $VOMC(0,1,1)$, respectively, and send their request to both $L_I(0,0)$ and $L_I(0,1)$ arbiters with weights 5 and 6, respectively. At $IM(1)$, $IP(1,0)$ selects $VOMC(1,0,0)$ and sends a request with weight 3 to $L_I(1,0)$ and $L_I(1,1)$ arbiters. IM link arbiters select the request with the largest weight and send the grant, as shown by the grant and accept steps in the figure. $L_I(0,0)$ selects the request with weight 6 from $IP(0,1)$ over the request with weight 5 from $IP(0,0)$, and $L_I(1,0)$ selects the request with weight 3 from $IP(1,0)$. In the second iteration, the same procedure is performed between unmatched inputs and unmatched output links. $L_I(0,1)$ is matched to $IP(0.0)$ in the second iteration, as indicated by the dashed line in the figure.

In Phase 2, link matching at CM is performed, as Figure 10.17 shows. In the request step, $VOMC(0,1,1)$, $VOMC(1,0,0)$, $VOMC(1,0,1)$, $VOMC(0,0,0)$, and $VOMC(0,0,1)$ send their request to their destined CM output-link arbiter $A_{CM}^E(r,j)$. In the grant step, $A_{CM}^E(0,1)$ receives two requests, and selects that from $A_{CM}^E(0,0)$ as this one has the larger weight. $A_{CM}^E(0,0)$, $A_{CM}^E(1,0)$, and $A_{CM}^E(1,1)$ receive a single request; therefore, the requests are granted. In the accept step, $A_{CM}^I(1,0)$ selects $A_{CM}^E(1,0)$, using the LQF policy, from the two grants received. $A_{CM}^I(0,0)$ accepts the single grant issued by $A_{CM}^E(0,1)$. $A_{CM}^I(0,1)$ accepts the single grant issued by $A_{CM}^E(0,0)$.

After Phases 1 and 2 are completed, input arbiters select the VOQ with cells for the matched OM with the largest occupancy.

This switch achieves 100% throughput under uniform traffic and above 99% throughput under unbalanced traffic. Moreover, these performance metrics are achieved without port matching and the buffers in OM use no speedup.

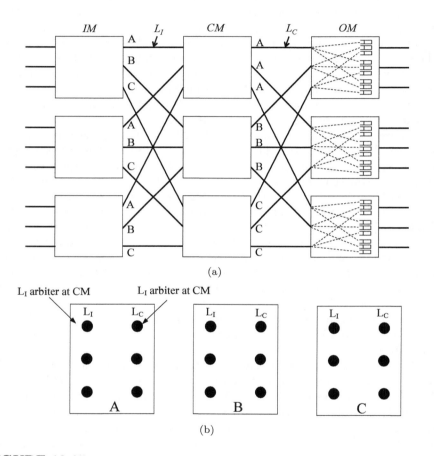

(a)

(b)

FIGURE 10.15
Example of matching process of the WCMM scheme in a ($n=m=k=3$) SSM
Clos-network switch.

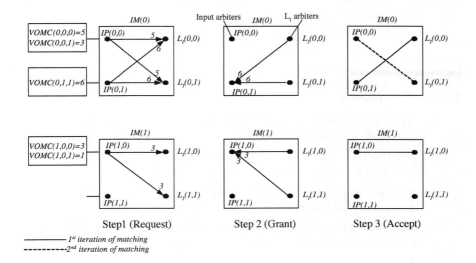

FIGURE 10.16
The example of matching within IM for the WCMM scheme in a $(n=m=k=2)$ SSM Clos-network switch.

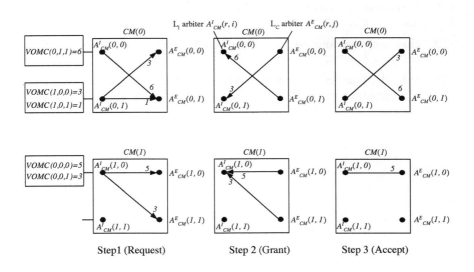

FIGURE 10.17
The example of matching within CM for the WCMM scheme in a $(n=m=k=2)$ SSM Clos-network switch.

This switch, as it has no buffers in CMs and uses FCFS selection in the OM buffers, forwards cells in sequence.

10.4 Exercises

1. What is the value m for a switch with $N = 1024$ and $n = 16$ to make a S^3 switch nonblocking?

2. What is the number of switch elements in an S^3 switch with $N = 2048$ and $n = 32$, if the switch is strictly nonblocking?

3. Show an example where an S^3 switch may have a larger number of crosspoint switches than a crossbar of the same size.

4. What is the speedup required by the IM memory in a MSM switch if the IMs have N VOQs in a $N = 4096$ and $k = 64$?

5. Discuss how the speedup of the memory in the MSM can be reduced.

6. Why does an MMM switch suffer from out-of-sequence cell forwarding?

7. Why might an MMM switch require a shorter configuration time than an S^3 switch?

8. How many queues are in an IM of an MM^eM switch with $N = 256$ and $n = k = 16$?

9. How many queues are in a CM of an MM^eM switch with $N = 256$ and $n = k = 16$?

10. Why does an SSM switch not need to perform port matching in its configuration scheme?

11

Buffer Management in Routers

CONTENTS

Buffer management is also important in the shared memory available in routers and hosts. The following are popular buffer management schemes used in shared memory. Some of these schemes may also be used in shared-memory switches.

11.1   Tail Drop

Tail drop is a simple buffer (queue) management scheme. It uses a threshold to decide when to drop a packet. Tail drop sets a threshold in terms of the number of packets and accepts all incoming packets until the threshold is exceeded. When this event occurs, subsequent incoming packets are dropped until the queue length decreases. Although the tail drop scheme is simple to implement and widely used in the Internet, it has some drawbacks such as monopolization, persistent queueing, burst losses, synchronization, and flow isolation (fairness).

Monopolization may occur when the packets of a single or a few flows monopolize the queue, leaving no room for other packets or flows. In such a scenario, packets of the flows that arrive after the monopolizing flow(s) are dropped according to a FIFO tail drop policy. This dropping action may cause large packet losses.

In general, queues must be large enough to store some number of packets corresponding to the bandwidth-delay product of the link. Although the desirable average queue lengths are small, the Transmission Control Protocol

(TCP) may account for large queueing delays even with nontransient traffic because TCP fills the queues until a loss occurs. This phenomenon is called persistent queueing and may lead to large queueing delays if the tail drop buffer management policy is employed in routers.

Actual network traffic is usually bursty, which means lots of packets are transmitted back-to-back in the network [104]. Many packets from the same flow may be lost if both the corresponding queue is full and the router adopts the tail drop policy. This is called burst losses and it may cause TCP to time out often and to retransmit them, which in turn results in a degraded throughput for those flows that have lost packets.

Synchronization may be another issue when the tail drop policy is used in routers. Synchronization may happen when multiple TCP flows go through the same router. After the queue of that router is filled up, the router starts dropping packets from all flows according to the tail drop policy. Then, the TCP flows that lose packets slow down and eventually the queue gets empty. After that, TCP at each host ramps up the transmission rate and the queue fills up quickly. The repetitive synchronized pattern arises because tail drop policy drops packets without differentiating any flow. During the ramp-up period, TCP's congestion windows of both senders are small and links become underutilized and the throughput decreases.

Another drawback of the tail drop policy is isolation, which occurs when noncooperative flow(s), such as User Datagram Protocol (UDP) flows with high sending rates share the queue with TCP flows. Because UDP does not use any congestion control mechanism, it may send high data rates, and this may be put TCP flows in disadvantage on sharing the same queue. This phenomenon may exacerbates when congestion occurs. Because tail drop policy drops packets regardless of the flow type, TCP flows may be unfairly penalized by the tail drop policy.

When the queue of a router is full, other packet-dropping policies, such as random drop on full and drop front on full, may be used. In drop front on full policy, the packet in the front of the queue is dropped to make room for the recently arrived packet [21], while random drop on full policy selects a random packet from the queue as a candidate and discards it to make space for the new packet. Those two dropping policies help TCP react to congestion faster than the regular tail drop policy, and in turn, to achieve a higher throughput.

11.2   Random Early Detection (RED)

Random early detection (RED) is an active queue management technique that allows routers to detect and indicate about an incipient congestion episode before the queue overflows [21, 40, 66]. This proactive approach probabilistically discards or marks the packets at routers as an indication of congestion. In this way, TCP can react to congestion before it gets severe. RED is mostly useful when the traffic is TCP-based because non-TCP traffic sources may not react to dropped/marked packets or these sources may immediately resend lost packets rather than slowing down, and the measure may not help in solving congestion [40]. The packet dropping or marking probability in RED is calculated based on the estimated value of the average queue length. Specifically, if the estimated average queue length is large, RED tends to drop/mark more packets than having a small estimated value for average queue length. Calculation of the average queue length follows a simple exponentially weighted moving average (avg), as in the (EWMA) formula [85], which is given as follows:

$$avg = avg + w(q - avg) \tag{11.1}$$

where $0 \leq w \leq 1$ and q is the recent queue length measure, which is updated after each packet arrival. Equation 11.1 works as a low-pass filter and considers the queue length history to calculate avg. EWMA is used to estimate the queue length because it prevents the queue length estimation from being affected by transient packet bursts. To update the estimation of the average queue length for an idle period (i.e., this is the period in which a queue does not have any packet), the update function used to calculate avg is:

$$avg = (1 - w)^m avg \tag{11.2}$$

where m denotes the number of small packets that could have been transmitted during that idle period. Here, m is calculated by extracting the current time from the time stamp indicating the beginning of the idle time, and dividing the result by the time to transmit the small packet. In this case, avg is updated at the end of an idle period if m packets had arrived to an empty queue during this idle period. After calculating the value at each packet arrival, avg is compared to two thresholds: a minimum threshold min_{th} and a maximum threshold max_{th}. If avg is smaller than min_{th}, none of the packets is dropped/marked by the router. If avg is larger than max_{th}, every packet is dropped/marked. Otherwise (i.e., if avg is between min_{th} and max_{th}) packets are dropped/marked with probability p_a, where p_a is a function of avg. Here p_a determines how frequent RED drops/marks a packet considering the extent of the congestion. Note that the probability of a packet belonging to a particular flow being dropped/marked is proportional to the bandwidth share of that flow. In other words, the more packets a flow sends to the queue, the more likely that these packets are dropped/marked if avg is set between min_{th} and max_{th}. For instance, a flow with a large sending rate would experience more packet drops than a flow with a small sending rate. This example shows that RED aims at penalizing flows in proportion to the amount of traffic they contribute. To calculate p_a, an auxiliary probability, p_b, is adopted as follows:

$$p_b = \frac{P_{max}(avg - min_{th})}{max_{th} - min_{th}} \tag{11.3}$$

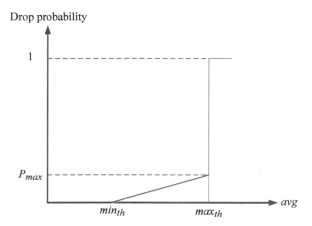

FIGURE 11.1
Packet dropping/marking probability of RED.

where p_{max} denotes the maximum value for p_b and it linearly increases when avg varies between max_{th} and min_{th}. Using p_b, p_a is calculated to determine the final dropping/marking probability for each packet as

$$p_a = \frac{p_b}{(1 - count)p_b} \qquad (11.4)$$

where *count* represents the number of packets in the queue after a packet is marked. Figure 11.1 shows how the dropping/marking probability of RED changes according to avg. Notice that dropping/marking probability is zero if avg is smaller than min_{th}, it is 1 if avg is larger than min_{th}, and it linearly increases if avg is between min_{th} and max_{th}.

RED can also estimate the average queue length in bytes in addition to the number of packets. In this operation mode, large packets are more likely to be dropped/marked than small packets, and the queue size more accurately reflects the average delay compared to the option where the number of packets is used for estimation of the average queue length. In this byte mode, RED follows:

$$p_b = \frac{p_b(\text{Packet size})}{\text{Maximum packet size}} \qquad (11.5)$$

For marking packets, rather than dropping them, RED sets a special field, called congestion experienced (CE) codepoint, in the IP header of packets [136]. These marked packets are then received by the destination, which in turn informs the sender about the incipient congestion at the router using acknowledgment (ACK) packets. This packet marking technique is called explicit congestion notification (ECN) and it is generally combined with RED. This combination can be useful when packet loss is undesirable. However, the use of ECN requires ECN-capable hosts and modifications to the transport protocols running at the hosts.

One of the main advantages of RED is that it does not require routers to keep per-flow state information. Therefore, this makes RED easy to implement with the combination of a FIFO scheduler to reduce the congestion in the network. RED might be convenient for the Internet backbone, where there may be hundreds of thousands of flows on a given link. RED also allows network operators to simultaneously achieve high throughput and low average delay [65]. However, RED is quite sensitive to the selection of min_{th}, max_{th}, and w, as they affect RED's performance. RED also requires some tuning of these parameters to adjust to current traffic conditions [65]. Moreover, RED does not consider packet priorities and it may drop/mark high-priority packets [40]. Another shortcoming of RED is the lack of fairness. RED does not always provide a fair share of the bandwidth among the flows sharing a link.

11.3   Weighted Random Early Detection (WRED)

Weighted RED (WRED) is a commercial (i.e., Cisco's) version of RED, which generally drops packets selectively based on Internet-Protocol (IP) precedence (i.e., using a 3-bit field in an IP packet header) [40]. WRED can selectively discard lower priority packets and provide differentiated services for different classes of flows when the interface of the router is congested. This means that low priority packets are more likely to be dropped/marked than the packets with high priority. For instance, a packet with an IP precedence value of 0 (i.e., the lowest priority) might have a minimum threshold of X packets, whereas a packet with an IP precedence of 1 might have a minimum threshold of $X + Y$ packets, where Y is a positive integer. Therefore, packets with IP precedence of 0 would be discarded before the packets with IP precedence of 1. WRED treats non-IP traffic with precedence 0 and thus non-IP traffic is more likely to be dropped than IP traffic. Figure 11.2 shows three different packet dropping/marking profiles for WRED. In this figure, MPD and AF stand for mark probability denominator and active flow, respectively. The mark probability denominator is used to calculate the packet dropping/marking probability as $1/MPD$, when the queue length reaches the maximum threshold, max_{th}. For instance, MPD is selected as 4 and this gives a 25% of dropping/marking probability when the average queue is at the maximum threshold. If max_{th} is exceeded, every packet is discarded regardless of its priority.

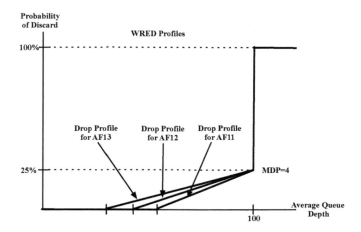

FIGURE 11.2
Three different packet dropping/marking profiles for WRED.

11.4   Fair Random Early Detection (FRED)

RED does not ensure a fair share of bandwidth for all flows as mentioned in [107]. Penalizing the flows in proportion to the amount of traffic they contribute does not always lead to a fair bandwidth sharing. Specifically, RED is not fair against low-speed TCP flows. For instance, if there are two TCP connections sharing one link unevenly, periodical packet dropping from the low-speed flow prevents the flow from obtaining a fair share, even if the high-speed flow experiences more packet drops [107]. Fair random early detection (FRED) addresses this unfairness problem and selectively sends feedback messages to a set of flows that have a large numbers of queued packets. FRED introduces two extra parameters, min_q and max_q, in addition to the parameters used in RED. Here min_q and max_q are the minimum and the maximum number of packets that each flow is allowed to buffer in the queue, respectively. FRED also introduces a global variable, $avgcq$, denoting an estimated value for the average per-flow queue size. Flows with fewer packets than $avgcq$ are favored over flows with more packets than $avgcq$. This value is calculated by dividing avg by the current number of active connections. Moreover, FRED maintains two extra variables specific to flow i, $qlen_i$, and $strike_i$, where i can be a number between zero and the current number of active flows (a flow is active if it has a packet queued in the buffer). Here, $qlen_i$ is the number of packets that flow i has currently in the queue and $strike_i$ is the number of times that flow i exceeds the max_q threshold. FRED does not allow flows with high strike values to queue more than $avgcq$ packets. This measure prevents flows from monopolizing the buffer space.

Each flow is allowed to send min_q packets without a loss. If the number of packets queued for flow i is smaller than min_q and the total queue length is less than max_{th}, an incoming packet for flow i is accepted. Here min_q value is set to 2 for small buffers and 4 for large buffers in FRED, because TCP cannot send more than three packets back-to-back (two because of the delayed acknowledgment, and an additional one because of a window increase). Furthermore, FRED gradually increases min_q to the average per-flow queue length ($avgcq$). However, FRED keeps a router from storing more than max_q packets of a flow.

While RED estimates the average queue length at each packet arrival, FRED calculates this estimate at both packet arrival and departure to reflect the queue variation more accurately. Moreover, this estimated value is not updated (i.e., when a packet is dropped) unless it is zero. Therefore, the same queue length can be measured if the input rate is essentially higher than the output link rate.

11.5 Adaptive Random Early Detection (ARED)

The average queueing delay and the throughput are sensitive to parameter selection and the traffic load under RED active queue management [109, 116, 132]. When the link is lightly congested and/or the maximum packet dropping probability (max_p) is high, the average queue size is expected to approach min_{th}, whereas when the link is more congested and/or max_p is low, the average queue size is expected to approach or exceed max_{th}. This means the average queueing delay is affected by the traffic load and the selection of RED's parameters. Furthermore, when the average queue length gets larger than max_{th}, RED does not perform well and the throughput decreases substantially [65]. ARED aims to solve these problems with minimal changes to the overall RED algorithm [63, 64]. The key idea in ARED is to keep the average queue length between min_{th} and max_{th} values by periodically adjusting the parameter max_p according to the observed traffic every 0.5 seconds. ARED uses a user-configurable variable, called $target$, to keep the average queue length within a target range halfway between min_{th} and max_{th}. Specifically, the target range is updated as

$$[min_{th} + 0.4 * (max_{th} - min_{th}), min_{th} + 0.6 * (max_{th} - min_{th})] \quad (11.6)$$

If ARED is too conservative, which means the average queue length oscillates around max_{th}, max_p is increased by a factor of α, where $\alpha \in (0.01, max_p/4)$. If ARED is too aggressive, which means the average queue length oscillates around min_{th}, max_p is decreased by multiplying it with a decrease factor, β, where β is selected as 0.9 in ARED's proposal. ARED also keeps max_p in the range of $[0.01, 0.5]$ to prevent the performance from degrading excessively. This guarantees the overall performance of RED to be acceptable during the transition period (i.e., the time needed for adapting to max_p updates) even though the average queue size might not be in its target range. Algorithm 2 shows ARED's pseudocode.

Algorithm 2 Adaptive RED Algorithm

Every *interval* second:
if $avg > target$ and $max_p \leq 0.5$ **then**
   increase max_p :
   $max_p \leftarrow max_p + \alpha$;
else
   if $avg < target$ and $max_p \leq 0.01$ **then**
      decrease max_p :
      $max_p \leftarrow max_p \times \beta$;
   end if
end if

Variables:
avg: average queue size

Fixed Parameters:
interval: time; 0.5 seconds
target: target for avg; $[min_{th} + 0.4 * (max_{th} - min_{th}), min_{th} + 0.6 * (max_{th} - min_{th})]$.
α : increment; $\min(0.01, max_p/4)$
β : decrease factor; 0.9

11.6 Differential Dropping (RIO)

Differential dropping or RED with In/Out bit (RIO) [41, 122] provides differentiated services to users by classifying their packets. RIO defines a service allocation profile for each user to tag users' packets as *In* or *Out*. Each packet entering the network is tagged according to the user profile and the congested routers through the network drop/mark *Out* packets with a higher probability than the probability that *In* packets have. The In/Out bit resembles the cell loss priority (CLP) bit of ATM networks and works in a similar way.

RIO uses two different packet dropping probabilities, called P_{in} and P_{out}, for *In* and *Out* packets, respectively. To determine P_{in} and P_{out}, routers calculate two estimated average queue lengths, avg_{in} and avg_{total}, every time a

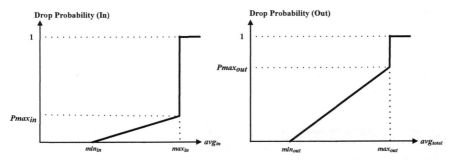

FIGURE 11.3
Packet dropping/marking probabilities for *In* and *Out* packets under different avg_{in} and avg_{total} values for RIO.

new packet arrives at the router. Upon a new *In* packet arrival, the router calculates avg_{in} by counting only the *In* packets in the queue. When an *Out* packet arrives, the router considers all the packets in the queue to calculate avg_{out} value. RIO-enabled routers differentiate the service they provide to *In* and *Out* packets by using different minimum and maximum threshold values, min_{in}, max_{in}, min_{out}, and max_{out}. Differentiation of *Out* packets is created by carefully selecting these parameters. First, min_{out} is selected to be smaller than min_{in}, so *Out* packets are discarded/marked earlier than *In* packets. Second, by selecting $max_{out} < max_{in}$, the router switches to the congestion control phase (i.e., the phase in which the router starts dropping every packet) from the congestion avoidance phase (i.e., probabilistic packet dropping phase) for *Out* packets earlier than *In* packets. Last, RIO drops/marks *Out* packets with a larger probability by setting $P_{max_{out}} > P_{max_{in}}$. Figure 11.3 shows how packet dropping/marking probabilities are changed for *In* and *Out* packets when avg_{in} and avg_{total} values vary, respectively. Algorithm 3 shows RIO's pseudocode.

11.7 Exercises

1. For a chain topology consisting of 2 hosts directly connected by a switch, what is the minimum queue size for each interface of the switch for a given bandwidth of 1 Gbps and a delay of 10 ms for each link in this chain topology? How many packets can be stored in the queue if the packet size is 1500 bytes?

2. Describe three drop modes of RED. (Hint: consider the cases $avg \leq min_{th}$, $min_{th} < avg \leq max_{th}$, and $max_{th} < avg$).

Algorithm 3 RIO Algorithm

For each packet arrival
if it is an In packet **then**
    calculate the average in queue size, avg_{in}
else
    calculate the average queue size, avg_{total}
end if

if it is an In packet **then**
    if $min_{in} \leq avg_{in} \leq max_{in}$ **then**
        calculate probability P_{in};
        with probability P_{in} drop this packet;
    else
        if $max_{in} \leq avg_{in}$ **then**
            drop this packet.
        end if
    end if
end if

if it is an Out packet **then**
    if $min_{out} \leq avg_{total} \leq max_{out}$ **then**
        calculate probability P_{out};
        with probability P_{out} drop this packet;
    else
        if $max_{out} \leq avg_{total}$ **then**
            drop this packet.
        end if
    end if
end if

---

3. A snapshot of a switch's queue employing RED is taken at time t. Calculate the final marking/dropping probability, p_a, of a packet received by the switch at time t for given min_{th}, max_{th}, avg, $count$, and P_{max} values as 5, 15, 8, 12 packets, and 0.02, respectively?

4. In one of the options of RED, the queue size is measured in bytes rather than in packets. Considering RED and the measure of the queue size in packets. Calculate the marking/dropping probability, p_a, of a 500-byte packet for the same parameters given in Exercise 3 and a maximal packet size of 1500 bytes.

5. List three major issues in RED and provide a possible solution for each of them.

6. Briefly explain why min_{out} is selected smaller than min_{in}, and max_{out} is selected as smaller than max_{in} in RED with In/Out bit (RIO)?

Part III

Data-Center Networks

12

Data Center Networks

CONTENTS

Data Center Networks (DCNs) may use some interconnection topologies that have their origins from computer architecture, but they have been adapted to communicate data packets between a large number of servers and other data center equipment. With the emergence of software-based networking, new architectures have been recently proposed. This chapter reviews some of these networks.

12.1 Introduction

Information is increasingly stored, and therefore accessed, from large data centers for technical and economical advantages. A data center can be simply described by large conglomerates of interconnected servers where many of them may be dedicated to perform a specific computing task or shared by a number of users for storage, processing, or access to large amounts of data. For a widespread and ubiquitous access to data by mobile or widely disseminated users, for example, data are increasingly stored in data centers. The flexible sharing of resources in a data center decreases investment costs and increases availability to scalable computing and communication resources. Data centers host, among other equipment, a very large number of servers. Examples of existing data centers are search engines, such as Google, Yahoo, and others, and those for cloud computing services, such as Amazon EC2 and others.

To make the communication among servers efficient, functional, and economically viable, it is important to provide a level of symmetry and modularity in a DCN. This network must provide high-performance communications for the large amounts of exchanged data in the data center.

This chapter describes several DCN architectures that have been recently introduced. There are several parameters that determine the features of DCNs, such as the number of servers that can be interconnected, the amount of network equipment needed, and other parameters that detail the performance features of the data center, such as bisection bandwidth and the largest distance between any two servers.

DCN architectures can be classified as switch-centric, server-centric, and hybrid structures. A switch-centric architecture uses switches to perform packet forwarding, whereas the server-centric architecture uses servers with multiple Network Interface Cards (NICs) to act as switches in addition to performing other computational functions. Hybrid architectures combine switches and servers for packet forwarding.

12.2   Switch-Centric Architectures

12.2.1   Three-Tier Network

The three-tier DCN architecture is considered a straightforward approach to building a DCN [1]. This architecture typically consists of three layers: access, aggregation, and core layers, as Figure 12.1 shows. In this network, servers are connected to the DCN through edge-level switches and placed in racks, in groups of 20 to 40. Each edge-level switch is connected to two aggregation-level switches for redundancy. These aggregation-level switches are further connected to core-level switches. Core switches serve as gateways and provide services such as firewall, load balancing, and Secure Socket Layer (SSL) offloading [1, 190]. The major advantage of this DCN is the simplicity of the topology at the expense of complex equipment and cabling. The major drawbacks of this architecture are the high cost, the low energy efficiency of the networking equipment, and the lack of agility and scalability [17].

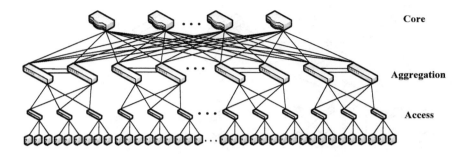

FIGURE 12.1
Three-tier DCN.

12.2.2 Fat-Tree Network

The fat-tree network is a highly scalable and cost-effective DCN architecture that aims to maximize end-to-end bisection bandwidth [4, 17]. A bisection is created by partitioning a network into two equally sized sets of nodes. The bandwidth of a bisection is found by summing all of the link capacities between two partitions and the smallest bandwidth of all those partitions is the bisection bandwidth [61]. The fat-tree network is a switch-centric architecture that can be built using commodity Gigabit Ethernet switches with the same number of ports to reduce the hardware cost. The size of the network is a function of the number of switch ports k. The network is formed by edge, aggregation, and core layers. In the edge layer, there are k pods, or groups of servers, each with $k^2/4$ servers. Figure 12.2 shows an example of a fat-tree network with four pods.

Each edge switch is directly connected to $k/2$ servers in a pod. The remaining $k/2$ ports of an edge switch are connected to $k/2$ aggregation switches. The total number of core switches in the DCN is $(k/2)^2$, and each of the core switches has one port connected to each of the k pods. A fat-tree network with k-port commodity switches can accommodate $k^3/4$ servers in total. One advantage of the fat-tree topology is that all switches are identical and possibly economical. This advantage may represent economic savings in equipment cost and a simplified architecture. Another advantage is the high fault tolerance provided by the use of multiple alternative paths between end nodes. A disadvantage of the fat-tree architecture is the use of large numbers of switches and high cabling costs [17].

12.2.3 VL2 Network

The VL2 network is a hierarchical fat-tree-based DCN architecture [17, 70]. Figure 12.3 shows a simple VL2 network. This switch-centric network uses three different types of switches: intermediate, aggregation, and top-of-rack

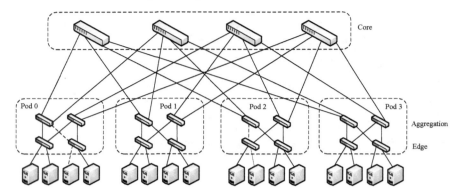

FIGURE 12.2
Fat-tree network with 4 pods and 16 servers.

(ToR) switches. The VL2 network targets the use of commodity switches. This network uses $D_A/2$ intermediate switches, D_I aggregation switches and $D_A D_I/4$ ToR switches. Intermediate and aggregation switches have a different number of ports, D_I and D_A, respectively. The number of servers in a VL2 network is $20(D_A D_I)/4$. VL2 also employs a load-balancing technique called valiant load balancing (VLB) to uniformly distribute the traffic among the network paths. One of the advantages of this architecture is its cost effectiveness due to the use of commodity switches throughout the network. Another advantage of VL2 is the ability to exploit the high bisection bandwidth because of the employed VLB technique.

12.3   Server-Centric Architectures

12.3.1   CamCube

The CamCube network is a server-centric architecture, proposed for building container-sized data centers [44, 71]. CamCube uses a 3D-Torus topology to directly interconnect the servers [181]. Figure 12.4 shows a 3D-Torus with 64 servers. CamCube, as a torus-based architecture, exploits network locality by placing the servers close to each other to increase communication efficiency. CamCube may reduce costs on network equipment (i.e., switches and/or routers) by using only servers to build the DCN. This approach may also reduce costs for cooling the network equipment. CamCube allows applications used in data centers to implement routing protocols through the CamCube Application Program Interface (API). The use of this application may result in achieving higher application-level performance. On the other

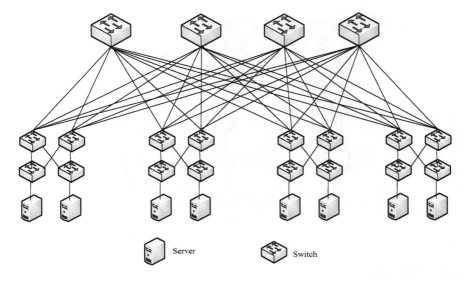

FIGURE 12.3
VL2 network.

hand, CamCube requires multiple NICs in each server to assemble a 3D Torus network. Furthermore, the use of the Torus topology by CamCube may result in long paths, or $O(N^{1/3})$, where N is the number of servers [181]. Furthermore, it has been claimed that routing complexity may be high [181].

12.4 Hybrid Architectures

12.4.1 DCell

The DCell network is a hybrid architecture; it uses switches and servers for packet forwarding, and it may be recursively scaled up to millions of servers [17, 72]. DCell uses a basic building block called $DCell_0$ to construct larger DCells (i.e., $DCell_1$, $DCell_2$, etc.). In general, $DCell_k$ ($k > 0$) is used to denote a level-k DCell that is constructed by combining $n + 1$ $DCell_{k-1}$s, where n denotes the number of servers in $DCell_0$. $DCell_0$ has n ($n \leq 8$) servers interconnected by a commodity switch. Moreover, each server in a $DCell_0$ is directly connected to a server in a different $DCell_0$. The interconnection of all $DCell_0$s forms a complete graph (i.e., every pair of $DCell_0$s in the network is interconnected) if each $DCell_0$ is considered as a large virtual node. Figure 12.5 shows a $DCell_1$, constructed with five $DCell_0$s and 4-port commodity switches.

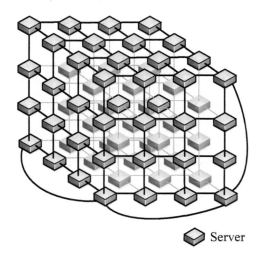

Server

FIGURE 12.4
3D-Torus with 64 servers.

The main advantage of DCell architecture is its high scalability, enabled by the recursive structure. DCell is also cost efficient because it uses commodity switches and servers to perform packet forwarding. The two main disadvantages of DCell are the long communication paths between two servers in the network and the additional NICs required for each server and the associated increased cabling costs.

12.4.2   BCube

BCube is another hybrid DCN architecture that can scale up through recursion [17, 71]. BCube employs servers and commodity switches as forwarding elements and is proposed for building so-called container data centers. The scalability of this structure is limited (up to thousands of servers) compared with fat-tree, VL2, and DCell. Conversely, BCube provides high bisection bandwidth and a graceful degradation of throughput under equipment failures [71]. As a recursive approach, BCube uses $BCube_0$ as a building block, which simply consists of n servers connected to an n-port switch. In BCube, n $BCube_0$s and n n-port switches comprise a $BCube_1$ network. In general, a $BCube_k$ ($k > 0$) is constructed by combining n $BCube_{k-1}$s and n^k n-port switches. In a $BCube_k$, there are n^{k+1} $k + 1$-port servers and $k + 1$ layers of switches. Figure 12.6(a) shows $BCube_1$ with $n = 4$ and Figure 12.6(b) shows a $BCube_k$ network. BCube is cost effective, has a high bisection bandwidth, and yields fault tolerance on equipment failures. However, BCube has limited scalability and its cabling cost is high because of the numerous interconnections

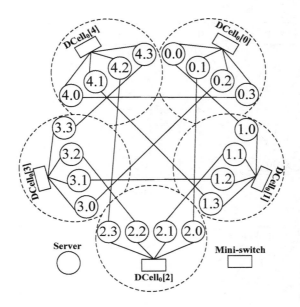

FIGURE 12.5
$DCell_1$ constructed with five $DCell_0$s ($n = 4$) [72].

among switches and servers. Furthermore, the number of NICs in a server in BCube is proportional to the depth of the network [17].

12.4.3 C-Through

C-Through [180] is a hybrid DCN architecture, which makes use of both electrical packet switching and optical circuit switching. C-Through combines optical circuit switching with traditional packet switching to make use of the bandwidth advantage that optical circuit switching has over (electrical) packet-switching technology. Other motivation for adopting this hybrid approach is the augmentation of both electrical and optical switching technologies to reduce network complexity, in terms of the number of links and the number of switches required [180]. The hybrid network of C-Through is referred to as hybrid packet and circuit (HyPaC) network and it consists of two parts: a tree-based three-tier electrical network, which ensures connectivity between each pair of ToR switches, and a reconfigurable optical network, which interconnects racks with high bandwidth optical links. Figure 12.7 shows the HyPaC network architecture with Core, Aggregation, and ToR switches, formed as a tree, with the electrical packet network at the top and the optical circuit-switching network with reconfigurable optical paths at the bottom. Due to the high maintenance cost of using a separate optical link between

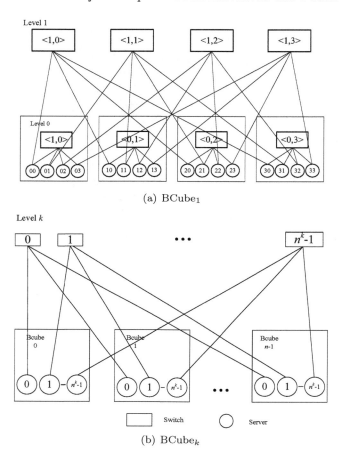

(a) BCube$_1$

(b) BCube$_k$

FIGURE 12.6
(a) *BCube$_1$* with $n = 4$. (b) *BCube$_k$* network.

each pair of racks, C-Through connects each rack to exactly one other rack at a time and it transiently changes the optical links between racks according to the traffic demand. The control plane of the system estimates the cross-rack traffic demand by observing the end-host socket buffer occupancies at runtime. C-Through isolates the two networks (i.e., electrical and optical) to de-multiplex traffic at either end-hosts or at the ToR switches. The idea behind isolating the traffic between circuit and packet-switched networks, and treating them as a single network, is to avoid a routing-based design, which may require switch modifications and the relative long convergence time of a routing protocol.

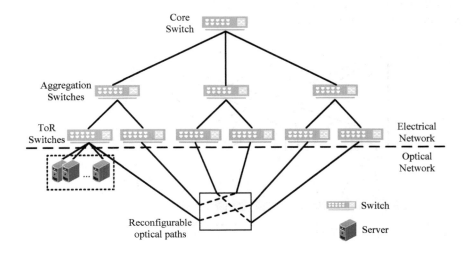

FIGURE 12.7
C-Through DCN.

12.4.4 Helios

Helios [60] is a hybrid DCN that uses both electrical packet switches and optical circuit switches to combine the benefits of both technologies. Helios may reduce the required number of switching elements, cabling, equipment cost, and power consumption as compared to other recently proposed DCN architectures by using this hybrid approach. Helios interconnects the so-called shipping containers, each equipped with networked servers and a built-in cooling system. Each container is referred to as a *pod* and it typically encloses between 250 to 1000 servers. Helios employs electrical packet switches because their configuration time is shorter than that of optical circuit switches. In addition, Helios uses electrical switches to distribute the bursty portion of traffic through the data center. Helios also employs optical circuit switches to provide reliable and high-bandwidth links between pods and to handle a portion of traffic between pods [60]. This DCN architecture has a two-tier multirooted tree architecture with pod (e.g., ToR) and core switches. Figure 12.8 shows a Helios network with two electrical packet switches, one micro-electromechanical system (MEMS)-based optical circuit switch at the core layer, and four electrical packet switches at the pod layer. Each electrical packet switch is equipped with optical transceivers and a multiplexer (MUX). Servers in Helios are connected to pod switches with short copper links. Half of the uplinks from each pod switch are connected to electrical packet switches at the core layer, using 10-Gbps fiber links, while the other half of those uplinks are connected to a single optical switch at the core layer by passing through a MUX. Each MUX combines two 10-Gbps fiber links and composes a link with 20-Gbps capacity, called *superlink*. Helios requires no modifications at the end-hosts and it only requires software modifications to switches.

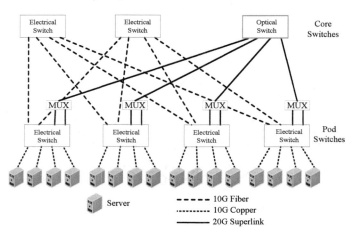

FIGURE 12.8
Helios DCN.

12.4.5   MDCube

MDCube [185] is a hierarchical DCN architecture that interconnects BCube-based containers for building mega data centers (i.e., from thousands to millions of servers). MDCube is designed to reduce the number of cables used in the interconnection of containers and to make use of high-speed (10 Gbps) interfaces of commodity on-the-shelf switches used in BCube. Fiber optic links are used to support hundreds of containers. Each container in MDCube has an ID that is mapped to a multidimensional tuple. Each container is connected to a neighbor container with a different tuple in one dimension. Figure 12.9 shows a two-dimensional MDCube network constructed by using nine $BCube_1$ containers. There are two types of links in MDCube: One is an intra-container link (represented by thin lines in Figure 12.9) to connect servers in each container, and the other one (represented by bold lines in Figure 12.9) is the high-speed inter-container link used in-between switches of different containers. The building process of a $(D+1)$ dimensional MDCube is as follows: The total number of containers in the system is denoted by $M = \prod_{d=0}^{D} m_d$, where m_d is the number of containers on dimension d. Each container has a container ID, cid, represented as $cid = c_D, c_{D-1}, \ldots, c_0$, where $c_d \in [0, m_d - 1]$ and $d \in [0, D]$. The number of switches in each container is $\sum_{d=0}^{D}(m_d - 1)$, where $(m_d - 1)$ is the number of the switches in dimension d. Each switch in MDCube is identified by its container and the switch ID in a BCube container as $\{cid, bwid\}$, where $cid \in [0, M - 1]$, $bwid \in [0, \sum_{d=0}^{D}(m_d - 1) - 1]$. Figure

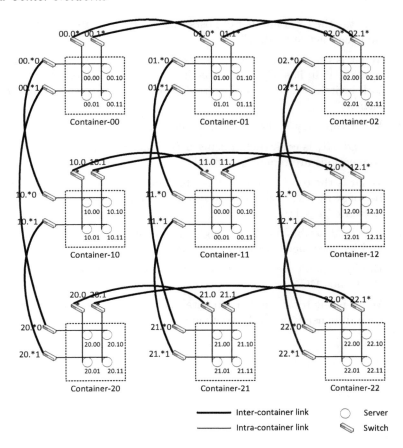

FIGURE 12.9
A 2-D MDCube constructed using $3x3 = 9$ $BCube_1$ containers.

12.9 shows an example of a 2-D MDCube built from nine $BCube_1$ containers with $n = 2$ (the number of ports for each switch in $BCube_1$) and $k = 1$ (the top-most level for $BCube_1$), where $D = 1$ and $M = 9$. Moreover, each switch ID in the figure is represented as c_0c_1, where $c_d \in [0, 2]$ and $bwid \in [0, 3]$.

12.5 Exercises

1. List the names of the layers forming the three-tier data center network architecture.

2. What is the total number of hosts that a fat-tree network with eight pods can accommodate?

3. What is the total number of hosts that a VL2 network can accommodate if it has $D_A/2$ intermediate, D_I aggregation, and $\frac{D_A D_I}{4}$ ToR switches (*Consider there are 20 servers connected to each ToR switch*)?

4. What topology is the CamCube network based on?

5. If a $DCell_1$ is constructed using 5 $DCell_0$, how many inter-$DCell_0$ connections are required in total?

6. What is the total number hosts, the number of levels, and the number of switches at each level, if n-port switches are used in a $BCube_k$ network ($n \geq 1$)?

7. Briefly describe the C-Through data center network architecture and list two merits of the architecture.

8. Briefly explain the role of transceivers and the multiplexers used in the Helios DCN.

9. What is the total number of containers, if there are 3 containers on each dimension of a 2-D MDCube DCN?

Bibliography

[1] Data Center Multi-Tier Model Design. CISCO Data Center Infrastructure 2.5 Design Guide, December 2007.

[2] BGP routing table analysis reports, January 2015. http://bgp.potaroo.net; accessed January, 2015.

[3] François Abel, Cyriel Minkenberg, Ronald P Luijten, Mitchell Gusat, and Ilias Iliadis. A four-terabit packet switch supporting long round-trip times. *IEEE Micro*, (1):10–24, 2003.

[4] Mohammad Al-Fares, Alexander Loukissas, and Amin Vahdat. A scalable, commodity data center network architecture. In *ACM SIGCOMM Computer Communication Review*, volume 38, pages 63–74. ACM, 2008.

[5] N.A. Al-Saber, S. Oberoi, T. Pedasanaganti, R. Rojas-Cessa, and S.G. Ziavras. Concatenating packets in variable-length input-queued packet switches with cell-based and packet-based scheduling. In *Sarnoff Symposium, 2008 IEEE*, pages 1–5, April 2008.

[6] Jose E. Moreira Alan Gara. Ibm blue gene supercomputer.

[7] James D Allen, Patrick T Gaughan, David E Schimmel, and Sudhakar Yalamanchili. Ariadne – an adaptive router for fault-tolerant multicomputers. *ACM SIGARCH Computer Architecture News*, 22(2):278–288, 1994.

[8] Thomas E Anderson, Susan S Owicki, James B Saxe, and Charles P Thacker. High-speed switch scheduling for local-area networks. *ACM Transactions on Computer Systems (TOCS)*, 11(4):319–352, 1993.

[9] Mutlu Arpaci and John A Copeland. Buffer management for shared-memory ATM switches. *Communications Surveys & Tutorials, IEEE*, 3(1):2–10, 2000.

[10] Florin Baboescu, Sumeet Singh, and George Varghese. Packet classification for core routers: Is there an alternative to CAMs? In *INFOCOM 2003. Twenty-Second Annual Joint Conference of the IEEE Computer and Communications. IEEE Societies*, volume 1, pages 53–63. IEEE, 2003.

[11] M. Bando and H.J. Chao. Flashtrie: Hash-based prefix-compressed trie for IP route lookup beyond 100 Gbps. pages 1–9, March 2010.

[12] L. Benini and G. De Micheli. Networks on chips: a new soc paradigm. *Computer*, 35(1):70–78, Jan 2002.

[13] Dimitri P Bertsekas, Robert G Gallager, and Pierre Humblet. *Data networks*, volume 2. Prentice-Hall International New Jersey, 1992.

[14] Laxmi N Bhuyan, Qing Yang, and Dharma P Agrawal. Performance of multiprocessor interconnection networks. *Computer*, 22(2):25–37, 1989.

[15] Chuan bi Lin and Roberto Rojas-Cessa. Minimizing scheduling complexity with a Clos-network space-space-memory (SSM) packet switch. In *High Performance Switching and Routing (HPSR), 2013 IEEE 14th International Conference on*, pages 15–20, July 2013.

[16] Andrea Bianco, M Franceschinis, S Ghisolfi, AM Hill, Emilio Leonardi, Fabio Neri, and R Webb. Frame-based matching algorithms for input-queued switches. In *High Performance Switching and Routing, 2002. Merging Optical and IP Technologies. Workshop on*, pages 69–76. IEEE, 2002.

[17] Kashif Bilal, Saif Ur Rehman Malik, Osman Khalid, Abdul Hameed, Enrique Alvarez, Vidura Wijaysekara, Rizwana Irfan, Sarjan Shrestha, Debjyoti Dwivedy, Mazhar Ali, et al. A taxonomy and survey on green data center networks. *Future Generation Computer Systems*, 36:189–208, 2014.

[18] Garrett Birkhoff. Tres observaciones sobre el algebra lineal. *Univ. Nac. Tucumán Rev. Ser. A*, 5:147–151, 1946.

[19] Burton H Bloom. Space/time trade-offs in hash coding with allowable errors. *Communications of the ACM*, 13(7):422–426, 1970.

[20] Rajendra V Boppana and Suresh Chalasani. A comparison of adaptive wormhole routing algorithms. *ACM SIGARCH Computer Architecture News*, 21(2):351–360, 1993.

[21] Bob Braden, David Clark, Jon Crowcroft, Bruce Davie, Steve Deering, Deborah Estrin, Sally Floyd, Van Jacobson, Greg Minshall, Craig Partridge, et al. Recommendations on queue management and congestion avoidance in the internet. *RFC 2309*, 1998.

[22] Eugene D Brooks. A butterfly processor-memory interconnection for a vector processing environment. *Parallel Computing*, 4(1):103–110, 1987.

[23] Lin Cai, Roberto Rojas-Cessa, and Taweesak Kijkanjanarat. Avoiding speedup from bandwidth overhead in a practical output-queued packet switch. In *Communications (ICC), 2011 IEEE International Conference on*, pages 1–5. IEEE, 2011.

[24] B. Cain and Kouvelas I. Fenner B. Thyagarajan A. Deering, S. RFC 3376 internet group management protocol, version 3. 2002.

[25] J.D. Carpineli and A.Y. Oruc. A nonbacktracking matrix decomposition algorithm for routing on Clos networks. *Communications, IEEE Transactions on*, 41(8):1245–1251, Aug 1993.

[26] Cheng-Shang Chang, Duan-Shin Lee, and Yi-Shean Jou. Load balanced birkhoff-von neumann switches. In *High Performance Switching and Routing, 2001 IEEE Workshop on*, pages 276–280. IEEE, 2001.

[27] Cheng-Shang Chang, Duan-Shin Lee, and Ching-Ming Lien. Load balanced birkhoff–von neumann switches, part ii: Multi-stage buffering. *Computer Communications*, 25(6):623–634, 2002.

[28] H. Jonathan Chao. Next generation routers. *Proceedings of the IEEE*, 90(9):1518–1558, Sep 2002.

[29] H Jonathan Chao and Bin Liu. *High performance switches and routers*. John Wiley & Sons, 2007.

[30] H Jonathan Chao and Jin-Soo Park. Centralized contention resolution schemes for a large-capacity optical ATM switch. In *ATM Workshop Proceedings, 1998 IEEE*, pages 11–16. IEEE, 1998.

[31] H Jonathan Chao, Jinsoo Park, Sertac Artan, Shi Jiang, and Guansong Zhang. Trueway: a highly scalable multi-plane multi-stage buffered packet switch. In *High Performance Switching and Routing, 2005. HPSR. 2005 Workshop on*, pages 246–253. IEEE, 2005.

[32] H.J. Chao, Zhigang Jing, and S.Y. Liew. Matching algorithms for three-stage bufferless Clos network switches. *Communications Magazine, IEEE*, 41(10):46–54, Oct 2003.

[33] X. Chen and J.F. Hayes. Call scheduling in multicasting packet switching. In *Communications, 1992. ICC '92, Conference record, SUPER-COMM/ICC '92, Discovering a New World of Communications., IEEE International Conference on*, pages 895–899 vol.2, Jun 1992.

[34] X. Cheng and I.F. Akyildiz. A finite buffer two class queue with different scheduling and push-out schemes. In *INFOCOM '92. Eleventh Annual Joint Conference of the IEEE Computer and Communications Societies, IEEE*, pages 231–241 vol.1, May 1992.

[35] Andrew A Chien and Jae H Kim. *Planar-adaptive routing: Low-cost adaptive networks for multiprocessors*, volume 20. ACM, 1992.

[36] Fabio M Chiussi, Joseph G Kneuer, and Vijay P Kumar. Low-cost scalable switching solutions for broadband networking: the atlanta architecture and chipset. *Communications Magazine, IEEE*, 35(12):44–53, 1997.

[37] A. Choudhury and E.L. Hahne. Dynamic queue length thresholds for shared-memory packet switches. *Networking, IEEE/ACM Transactions on*, 6(2):130–140, Apr 1998.

[38] Shang-Tse Chuang, Ashish Goel, Nick McKeown, and Balaji Prabhakar. Matching output queueing with a combined input/output-queued switch. *Selected Areas in Communications, IEEE Journal on*, 17(6):1030–1039, 1999.

[39] I. Cidon, L. Georgiadis, R. Guerin, and A. Khamisy. Optimal buffer sharing. *Selected Areas in Communications, IEEE Journal on*, 13(7):1229–1240, Sep 1995.

[40] CISCO. Congestion avoidance overview. Cisco IOS Quality of Service Solutions Configuration Guide Release 12.2.

[41] David Clark and John Wroclawski. An approach to service allocation in the internet. Technical report, Internet Draft draft-clark-diff-svc-alloc-00. txt, July 1997, also talk by D. Clark in the Int-Serv WG at the Munich IETF, 1997.

[42] Charles Clos. A study of non-blocking switching networks. *Bell System Technical Journal*, 32(2):406–424, 1953.

[43] Thomas H Cormen, Charles E Leiserson, Ronald L Rivest, Clifford Stein, et al. *Introduction to algorithms*, volume 2. MIT press Cambridge, 2001.

[44] P Costa, A Donnelly, G Oshea, and A Rowstron. CamCube: a key-based data center. Technical report, Technical Report MSR TR-2010-74, Microsoft Research, 2010.

[45] JG Dai and Balaji Prabhakar. The throughput of data switches with and without speedup. In *INFOCOM 2000. Nineteenth Annual Joint Conference of the IEEE Computer and Communications Societies. Proceedings. IEEE*, volume 2, pages 556–564. IEEE, 2000.

[46] William J. Dally and Hiromichi Aoki. Deadlock-free adaptive routing in multicomputer networks using virtual channels. *Parallel and Distributed Systems, IEEE Transactions on*, 4(4):466–475, 1993.

[47] William James Dally and Brian Patrick Towles. *Principles and practices of interconnection networks*. Elsevier, 2004.

[48] Víctor López de Buen. Multistage interconnection networks in multiprocessor systems: a simulation study. *Questiió: Quaderns d'Estadística, Sistemes, Informatica i Investigació Operativa*, 11(3):73–86, 1987.

[49] Mikael Degermark, Andrej Brodnik, Svante Carlsson, and Stephen Pink. Small forwarding tables for fast routing lookups. In *Proceedings of the*

ACM SIGCOMM '97 Conference on Applications, Technologies, Architectures, and Protocols for Computer Communication, SIGCOMM '97, pages 3–14, New York, NY, USA, 1997. ACM.

[50] Michel Devault, J-Y Cochennec, and Michel Servel. The'prelude'atd experiment: assessments and future prospects. *Selected Areas in Communications, IEEE Journal on*, 6(9):1528–1537, 1988.

[51] Ziqian Dong and Roberto Rojas-Cessa. Long round-trip time support with shared-memory crosspoint buffered packet switch. In *High Performance Interconnects, 2005. Proceedings. 13th Symposium on*, pages 138–143. IEEE, 2005.

[52] Ziqian Dong and Roberto Rojas-Cessa. Shared-memory combined input-crosspoint buffered packet switch for differentiated services. *2006 IEEE Global Conference on Telecommunications*, pages 1–6, 2006.

[53] Ziqian Dong and Roberto Rojas-Cessa. Input-and output-based shared-memory crosspoint-buffered packet switches for multicast traffic switching and replication. In *Communications, 2008. ICC'08. IEEE International Conference on*, pages 5659–5663. IEEE, 2008.

[54] Ziqian Dong and Roberto Rojas-Cessa. Non-blocking memory-memory-memory Clos-network packet switch. In *Sarnoff Symposium, 2011 34th IEEE*, pages 1–5. IEEE, 2011.

[55] Ziqian Dong, Roberto Rojas-Cessa, and Eiji Oki. Buffered Clos-network packet switch with per-output flow queues. *Electronics letters*, 47(1):32–34, 2011.

[56] Ran Duan and Seth Pettie. Approximating maximum weight matching in near-linear time. In *Foundations of Computer Science (FOCS), 2010 51st Annual IEEE Symposium on*, pages 673–682. IEEE, 2010.

[57] Jose Duato, Sudhakar Yalamanchili, and Lionel M Ni. *Interconnection networks: An engineering approach*. Morgan Kaufmann, 2003.

[58] Itamar Elhanany and Mounir Hamdi. *High-performance packet switching architectures*. Springer, 2007.

[59] N. Endo, T. Kozaki, T. Ohuchi, H. Kuwahara, and S. Gohara. Shared buffer memory switch for an ATM exchange. *Communications, IEEE Transactions on*, 41(1):237–245, Jan 1993.

[60] Nathan Farrington, George Porter, Sivasankar Radhakrishnan, Hamid Hajabdolali Bazzaz, Vikram Subramanya, Yeshaiahu Fainman, George Papen, and Amin Vahdat. Helios: a hybrid electrical/optical switch architecture for modular data centers. *ACM SIGCOMM Computer Communication Review*, 41(4):339–350, 2011.

[61] Nathan Farrington, Erik Rubow, and Amin Vahdat. Data center switch architecture in the age of merchant silicon. In *High Performance Interconnects, 2009. HOTI 2009. 17th IEEE Symposium on*, pages 93–102. IEEE, 2009.

[62] Sergio A Felperin, Luis Gravano, Gustavo D Pifarré, and Jorge LC Sanz. Fully-adaptive routing: packet switching performance and wormhole algorithms. In *Proceedings of the 1991 ACM/IEEE conference on Supercomputing*, pages 654–663. ACM, 1991.

[63] Wu-chang Feng, Dilip D Kandlur, Debanjan Saha, and Kang G Shin. Techniques for eliminating packet loss in congested TCP/IP networks. *Ann Arbor*, 1001:63130, 1997.

[64] Wu-chang Feng, Dilip D Kandlur, Debanjan Saha, and Kang G Shin. A self-configuring RED gateway. In *INFOCOM'99. Eighteenth Annual Joint Conference of the IEEE Computer and Communications Societies. Proceedings. IEEE*, volume 3, pages 1320–1328. IEEE, 1999.

[65] Sally Floyd, Ramakrishna Gummadi, Scott Shenker, et al. Adaptive RED: An algorithm for increasing the robustness of RED's active queue management, 2001.

[66] Sally Floyd and Van Jacobson. Random early detection gateways for congestion avoidance. *Networking, IEEE/ACM Transactions on*, 1(4):397–413, 1993.

[67] Vince Fuller, Tony Li, Jessica Yu, and Kannan Varadhan. Classless inter-domain routing (CIDR): an address assignment and aggregation strategy. Technical report, 1993.

[68] P. Giaccone, B. Prabhakar, and D. Shah. Randomized scheduling algorithms for high-aggregate bandwidth switches. *Selected Areas in Communications, IEEE Journal on*, 21(4):546–559, May 2003.

[69] James R. Goodman and Carlo H. Sequin. Hypertree: A multiprocessor interconnection topology. *Computers, IEEE Transactions on*, 100(12):923–933, 1981.

[70] Albert Greenberg, James R Hamilton, Navendu Jain, Srikanth Kandula, Changhoon Kim, Parantap Lahiri, David A Maltz, Parveen Patel, and Sudipta Sengupta. VL2: a scalable and flexible data center network. In *ACM SIGCOMM Computer Communication Review*, volume 39, pages 51–62. ACM, 2009.

[71] Chuanxiong Guo, Guohan Lu, Dan Li, Haitao Wu, Xuan Zhang, Yunfeng Shi, Chen Tian, Yongguang Zhang, and Songwu Lu. BCube: a high performance, server-centric network architecture for modular data centers. *ACM SIGCOMM Computer Communication Review*, 39(4):63–74, 2009.

[72] Chuanxiong Guo, Haitao Wu, Kun Tan, Lei Shi, Yongguang Zhang, and Songwu Lu. Dcell: a scalable and fault-tolerant network structure for data centers. *ACM SIGCOMM Computer Communication Review*, 38(4):75–86, 2008.

[73] Anil K Gupta, Luis Orozco Barbosa, and ND Georganas. 16× 16 limited intermediate buffer switch module for ATM networks. In *Global Telecommunications Conference, 1991. GLOBECOM'91.'Countdown to the New Millennium. Featuring a Mini-Theme on: Personal Communications Services*, pages 939–943. IEEE, 1991.

[74] Anil K Gupta, Luis Orozco Barbosa, and ND Georganas. Limited intermediate buffer switch modules and their interconnection networks for B-ISDN. In *Communications, 1992. ICC'92, Conference record, SUPERCOMM/ICC'92, Discovering a New World of Communications., IEEE International Conference on*, pages 1646–1650. IEEE, 1992.

[75] Pankaj Gupta, Steven Lin, and Nick McKeown. Routing lookups in hardware at memory access speeds. In *INFOCOM'98. Seventeenth Annual Joint Conference of the IEEE Computer and Communications Societies. Proceedings. IEEE*, volume 3, pages 1240–1247. IEEE, 1998.

[76] Pankaj Gupta and Nick McKeown. Packet classification using hierarchical intelligent cuttings. In *Hot Interconnects VII*, pages 34–41, 1999.

[77] Pankaj Gupta and Nick McKeown. Algorithms for packet classification. *Network, IEEE*, 15(2):24–32, 2001.

[78] Dan Gusfield and Robert W Irving. *The stable marriage problem: structure and algorithms*, volume 54. MIT press Cambridge, 1989.

[79] John L Hennessy and David A Patterson. *Computer architecture: a quantitative approach*. Elsevier, 2009.

[80] M.G. Hluchyj and M.J. Karol. Queueing in high-performance packet switching. *Selected Areas in Communications, IEEE Journal on*, 6(9):1587–1597, Dec 1988.

[81] John E Hopcroft and Richard M Karp. An $n^{5/2}$ algorithm for maximum matchings in bipartite graphs. *SIAM Journal on computing*, 2(4):225–231, 1973.

[82] J.Y. Hui and T. Renner. Queueing strategies for multicast packet switching. In *Global Telecommunications Conference, 1990, and Exhibition. 'Communications: Connecting the Future', GLOBECOM '90., IEEE*, pages 1431–1437 vol.3, Dec 1990.

[83] M. Irland. Buffer management in a packet switch. *Communications, IEEE Transactions on*, 26(3):328–337, Mar 1978.

[84] Sundar Iyer and Nick McKeown. Making parallel packet switches practical. In *INFOCOM 2001. Twentieth Annual Joint Conference of the IEEE Computer and Communications Societies. Proceedings. IEEE*, volume 3, pages 1680–1687. IEEE, 2001.

[85] Van Jacobson. Congestion avoidance and control. *ACM SIGCOMM Computer Communication Review*, 18(4):314–329, 1988.

[86] Tara Javidi, Robert Magill, and Terry Hrabik. A high-throughput scheduling algorithm for a buffered crossbar switch fabric. In *Communications, 2001. ICC 2001. IEEE International Conference on*, volume 5, pages 1586–1591. IEEE, 2001.

[87] F. Kamoun and L. Kleinrock. Analysis of shared finite storage in a computer network node environment under general traffic conditions. *Communications, IEEE Transactions on*, 28(7):992–1003, Jul 1980.

[88] Yossi Kanizo, David Hay, and Isaac Keslassy. The crosspoint-queued switch. In *INFOCOM 2009, IEEE*, pages 729–737. IEEE, 2009.

[89] Mark J Karol, Michael G Hluchyj, and Samuel P Morgan. Input versus output queueing on a space-division packet switch. *Communications, IEEE Transactions on*, 35(12):1347–1356, 1987.

[90] Manolis Katevenis and Georgios Passas. Variable-size multipacket segments in buffered crossbar (cicq) architectures. In *Communications, 2005. ICC 2005. 2005 IEEE International Conference on*, volume 2, pages 999–1004. IEEE, 2005.

[91] Manolis Katevenis, Giorgos Passas, Dimitrios Simos, Ioannis Papaefstathiou, and Nikolaos Chrysos. Variable packet size buffered crossbar (CICQ) switches. In *Communications, 2004 IEEE International Conference on*, volume 2, pages 1090–1096. IEEE, 2004.

[92] Isaac Keslassy and Nick McKeown. Maintaining packet order in two-stage switches. In *INFOCOM 2002. Twenty-First Annual Joint Conference of the IEEE Computer and Communications Societies. Proceedings. IEEE*, volume 2, pages 1032–1041. IEEE, 2002.

[93] Leonard Kleinrock. *Theory, Volume 1, Queueing Systems*. Wiley-Interscience, 1975.

[94] T. Kozaki, Y. Sakurai, O. Matsubara, M. Mizukami, M. Uchida, Y. Sato, and K. Asano. 32×32 shared buffer type ATM switch vlsis for b-isdn. In *Communications, 1991. ICC '91, Conference Record. IEEE International Conference on*, pages 711–715 vol.2, Jun 1991.

[95] Pattabhiraman Krishna, Naimish S Patel, Anna Charny, and Robert J Simcoe. On the speedup required for work-conserving crossbar switches.

Selected Areas in Communications, IEEE Journal on, 17(6):1057–1066, 1999.

[96] HT Kung, Trevor Blackwell, and Alan Chapman. Credit-based flow control for ATM networks: credit update protocol, adaptive credit allocation and statistical multiplexing. In *ACM SIGCOMM Computer Communication Review*, volume 24, pages 101–114. ACM, 1994.

[97] HT Kung and Koling Chang. Receiver-oriented adaptive buffer allocation in credit-based flow control for ATM networks. In *INFO-COM'95. Fourteenth Annual Joint Conference of the IEEE Computer and Communications Societies. Bringing Information to People. Proceedings. IEEE*, pages 239–252. IEEE, 1995.

[98] H. Kuwahara, N. Endo, M. Ogino, T. Kozaki, Y. Sakurai, and S. Gohara. A shared buffer memory switch for an atm exchange. In *Communications, 1989. ICC '89, BOSTONICC/89. Conference record. 'World Prosperity Through Communications', IEEE International Conference on*, pages 118–122, Jun 1989.

[99] Duncan H Lawrie. Access and alignment of data in an array processor. *Computers, IEEE Transactions on*, 100(12):1145–1155, 1975.

[100] Hyun Yeop Lee, F.K. Hwang, and J.D. Carpinelli. A new decomposition algorithm for rearrangeable Clos interconnection networks. *Communications, IEEE Transactions on*, 44(11):1572–1578, Nov 1996.

[101] Tony T Lee and Cheuk H Lam. Path switching-a quasi-static routing scheme for large-scale ATM packet switches. *Selected Areas in Communications, IEEE Journal on*, 15(5):914–924, 1997.

[102] Charles E Leiserson. Fat-trees: universal networks for hardware-efficient supercomputing. *Computers, IEEE Transactions on*, 100(10):892–901, 1985.

[103] Charles E Leiserson, Zahi S Abuhamdeh, David C Douglas, Carl R Feynman, Mahesh N Ganmukhi, Jeffrey V Hill, Daniel Hillis, Bradley C Kuszmaul, Margaret A St Pierre, David S Wells, et al. The network architecture of the connection machine cm-5. In *Proceedings of the fourth annual ACM symposium on Parallel algorithms and architectures*, pages 272–285. ACM, 1992.

[104] Will E Leland, Murad S Taqqu, Walter Willinger, and Daniel V Wilson. On the self-similar nature of Ethernet traffic (extended version). *Networking, IEEE/ACM Transactions on*, 2(1):1–15, 1994.

[105] Yihan Li, Shivendra Panwar, and H Jonathan Chao. The dual round robin matching switch with exhaustive service. In *High Performance Switching and Routing, 2002. Merging Optical and IP Technologies. Workshop on*, pages 58–63. IEEE, 2002.

[106] Chuan-Bi Lin and Roberto Rojas-Cessa. Frame occupancy-based dispatching schemes for buffered three-stage Clos-network switches. In *Networks, 2005. Jointly held with the 2005 IEEE 7th Malaysia International Conference on Communication., 2005 13th IEEE International Conference on*, volume 2, pages 5–pp. IEEE, 2005.

[107] Dong Lin and Robert Morris. Dynamics of random early detection. *ACM SIGCOMM Computer Communication Review*, 27(4):127–137, 1997.

[108] Daniel H. Linder and James C. Harden. An adaptive and fault tolerant wormhole routing strategy for k-ary n-cubes. *Computers, IEEE Transactions on*, 40(1):2–12, 1991.

[109] Martin May, Jean Bolot, Christophe Diot, and Bryan Lyles. Reasons not to deploy RED. In *Quality of Service, 1999. IWQoS'99. 1999 Seventh International Workshop on*, pages 260–262. IEEE, 1999.

[110] Nicholas William McKeown. *Scheduling algorithms for input-queued cell switches*. PhD thesis, Citeseer, 1992.

[111] Nick McKeown. The *i*SLIP scheduling algorithm for input-queued switches. *Networking, IEEE/ACM Transactions on*, 7(2):188–201, 1999.

[112] Nick McKeown, Martin Izzard, Adisak Mekkittikul, William Ellersick, and Mark Horowitz. Tiny tera: a packet switch core. *Micro, IEEE*, 17(1):26–33, 1997.

[113] Nick McKeown, Adisak Mekkittikul, Venkat Anantharam, and Jean Walrand. Achieving 100% throughput in an input-queued switch. *Communications, IEEE Transactions on*, 47(8):1260–1267, 1999.

[114] L. Mhamdi and M. Hamdi. MCBF: a high-performance scheduling algorithm for buffered crossbar switches. *Communications Letters, IEEE*, 7(9):451–453, Sept 2003.

[115] Steven E Minzer. Broadband ISDN and asynchronous transfer mode (ATM). *Communications Magazine, IEEE*, 27(9):17–24, 1989.

[116] Vishal Misra, Wei-Bo Gong, and Don Towsley. Fluid-based analysis of a network of AQM routers supporting tcp flows with an application to RED. *ACM SIGCOMM Computer Communication Review*, 30(4):151–160, 2000.

[117] Donald R Morrison. PATRICIA– practical algorithm to retrieve information coded in alphanumeric. *Journal of the ACM (JACM)*, 15(4):514–534, 1968.

[118] J. Mun and H. Lim. New approach for efficient IP address lookup using a Bloom filter in trie-based algorithms. *Computers, IEEE Transactions on*, PP(99):1–1, 2015.

[119] Masayoshi Nabeshima. Performance evaluation of a combined input-and crosspoint-queued switch. *IEICE Transactions on Communications*, 83(3):737–741, 2000.

[120] Ted Nesson and S Lennart Johnsson. ROMM routing on mesh and torus networks. In *Proceedings of the seventh annual ACM symposium on Parallel algorithms and architectures*, pages 275–287. ACM, 1995.

[121] John Y Ngai and Charles L Seitz. A framework for adaptive routing in multicomputer networks. *ACM SIGARCH Computer Architecture News*, 19(1):6–14, 1991.

[122] Kathleen Nichols, Van Jacobson, and L Zhang. A two-bit differentiated services architecture for the Internet. *Request for Comments: 2638*, 1999.

[123] Gerd Niestegge. The leaky bucketpolicing method in the ATM (Asynchronous Transfer Mode) network. *International Journal of Digital & Analog Communication Systems*, 3(2):187–197, 1990.

[124] Stefan Nilsson and Gunnar Karlsson. IP-address lookup using LC-tries. *Selected Areas in Communications, IEEE Journal on*, 17(6):1083–1092, 1999.

[125] Satoshi Nojima, Eiichi Tsutsui, Haruki Fukuda, and Masamichi Hashimoto. Integrated services packet network using bus matrix switch. *Selected Areas in Communications, IEEE Journal on*, 5(8):1284–1292, 1987.

[126] W. Noureddine and F. Tobagi. Selective back-pressure in switched ethernet LANs. In *Global Telecommunications Conference, 1999. GLOBECOM '99*, volume 2, pages 1256–1263 vol.2, 1999.

[127] Eiji Oki, Zhigang Jing, Roberto Rojas-Cessa, and H Jonathan Chao. Concurrent round-robin dispatching scheme in a Clos-network switch. In *ICC*, volume 1, pages 107–111, 2001.

[128] Eiji Oki, Zhigang Jing, Roberto Rojas-Cessa, and H Jonathan Chao. Concurrent round-robin-based dispatching schemes for Clos-network switches. *Networking, IEEE/ACM Transactions on*, 10(6):830–844, 2002.

[129] Eiji Oki, Roberto Rojas-Cessa, and H Jonathan Chao. A pipeline-based approach for maximal-sized matching scheduling in input-buffered switches. *Communications Letters, IEEE*, 5(6):263–265, 2001.

[130] Eiji Oki and Naoaki Yamanaka. Scalable crosspoint buffering ATM switch architecture using distributed arbitration scheme. In *IEEE ATM Workshop 1997. Proceedings*, pages 28–35. IEEE, 1997.

[131] Eiji Oki and Naoaki Yamanaka. Tandem-crosspoint ATM switch with input and output buffers. *Communications Letters, IEEE*, 2(7):189–191, 1998.

[132] Teunis J Ott, TV Lakshman, and Larry H Wong. Sred: stabilized RED. In *INFOCOM'99. Eighteenth Annual Joint Conference of the IEEE Computer and Communications Societies. Proceedings. IEEE*, volume 3, pages 1346–1355. IEEE, 1999.

[133] Rina Panigrahy and Samar Sharma. Reducing TCAM power consumption and increasing throughput. In *High Performance Interconnects, 2002. Proceedings. 10th Symposium on*, pages 107–112. IEEE, 2002.

[134] Balaji Prabhakar and Nick McKeown. On the speedup required for combined input-and output-queued switching. *Automatica*, 35(12):1909–1920, 1999.

[135] Y Awdeh Ra'ed and Hussein T Mouftah. Survey of ATM switch architectures. *Computer networks and ISDN systems*, 27(12):1567–1613, 1995.

[136] K Ramakrishnan, Sally Floyd, David Black, et al. The addition of explicit congestion notification (ECN) to IP, 2001.

[137] Erwin P Rathgeb, Thomas H Theimer, and Manfred N Huber. Buffering concepts for ATM switching networks. In *Global Telecommunications Conference, 1988, and Exhibition.'Communications for the Information Age.'Conference Record, GLOBECOM'88., IEEE*, pages 1277–1281. IEEE, 1988.

[138] Enrico Del Re and Romano Fantacci. Performance evaluation of input and output queueing techniques in ATM switching systems. *Communications, IEEE Transactions on*, 41(10):1565–1575, 1993.

[139] R. Rojas-Cessa, E. Oki, Z. Jing, and H.J. Chao. CIXB-1: combined input-one-cell-crosspoint buffered switch. In *High Performance Switching and Routing, 2001 IEEE Workshop on*, pages 324–329, 2001.

[140] R. Rojas-Cessa, L. Ramesh, Z. Dong, L. Cai, and N. Ansari. Parallel-search trie-based scheme for fast IP lookup. pages 210–214, November 2007.

[141] Roberto Rojas-Cessa and Ziqian Dong. Combined input-crosspoint buffered packet switch with shared crosspoint buffers. *Proceedings of the 39th Conference on Information Sciences and Systems*, pages 16–18, 2005.

[142] Roberto Rojas-Cessa and Ziqian Dong. Combined input-crosspoint buffered packet switch with flexible access to crosspoints buffers. In

Devices, Circuits and Systems, Proceedings of the 6th International Caribbean Conference on, pages 255–260. IEEE, 2006.

[143] Roberto Rojas-Cessa and Ziqian Dong. Load-balanced combined input-crosspoint buffered packet switches. *Communications, IEEE Transactions on*, 59(5):1421–1433, 2011.

[144] Roberto Rojas-Cessa, Ziqian Dong, and Zhen Guo. Load-balanced combined input-crosspoint buffered packet switch and long round-trip times. *Communications Letters, IEEE*, 9(7):661–663, 2005.

[145] Roberto Rojas-Cessa, Taweesak Kijkanjanarat, Wara Wangchai, Krutika Patil, and Narathip Thirapittayatakul. Helix: Ip lookup scheme based on helicoidal properties of binary trees. *Computer Networks*, 89:78–89, 2015.

[146] Roberto Rojas-Cessa and Chuan-bi Lin. Captured-frame eligibility and round-robin matching for input-queued packet switches. *Communications Letters, IEEE*, 8(9):585–587, 2004.

[147] Roberto Rojas-Cessa and Chuan-Bi Lin. Captured-frame matching schemes for scalable input-queued packet switches. *Computer communications*, 30(10):2149–2161, 2007.

[148] Roberto Rojas-Cessa, Chuan-Bi Lin, and Ziqian Dong. Space-space-memory (SSM) Clos-network packet switch, March 31 2015. US Patent 8,995,456.

[149] Roberto Rojas-Cessa and Eiji Oki. Round-robin selection with adaptable-size frame in a combined input-crosspoint buffered switch. *Communications Letters, IEEE*, 7(11):555–557, 2003.

[150] Roberto Rojas-Cessa, Eiji Oki, and H Jonathan Chao. CIXOB-k: Combined input-crosspoint-output buffered packet switch. In *Global Telecommunications Conference, 2001. GLOBECOM'01. IEEE*, volume 4, pages 2654–2660. IEEE, 2001.

[151] Roberto Rojas-Cessa, Eiji Oki, and H Jonathan Chao. CIXOB-k: Combined input-crosspoint-output buffered packet switch. In *Global Telecommunications Conference, 2001. GLOBECOM'01. IEEE*, volume 4, pages 2654–2660. IEEE, 2001.

[152] Miguel Ruiz-Sanchez, Ernst W Biersack, Walid Dabbous, et al. Survey and taxonomy of ip address lookup algorithms. *Network, IEEE*, 15(2):8–23, 2001.

[153] Devavrat Shah and Pankaj Gupta. Fast incremental updates on ternary-cams for routing lookups and packet classification. In *Proc. of Hot Interconnects-8, Stanford, CA, USA*, 2000.

[154] Devavrat Shah and Pankaj Gupta. Fast updating algorithms for TCAM. *Micro, IEEE*, 21(1):36–47, 2001.

[155] Howard Jay Siegel, Wayne G Nation, Clyde P Kruskal, and Leonard M Napolitano. Using the multistage cube network topology in parallel supercomputers. *Proceedings of the IEEE*, 77(12):1932–1953, 1989.

[156] Howard Jay Siegel, Thomas Schwederski, David G Meyer, and William Tsun-yuk Hsu. Large-scale parallel processing systems. *Microprocessors and Microsystems*, 11(1):3–20, 1987.

[157] Arjun Singh, William J Dally, Brian Towles, and Amit K Gupta. Locality-preserving randomized oblivious routing on torus networks. In *Proceedings of the fourteenth annual ACM symposium on Parallel algorithms and architectures*, pages 9–13. ACM, 2002.

[158] Ed Spitznagel, David Taylor, and Jonathan Turner. Packet classification using extended TCAMs. In *Network Protocols, 2003. Proceedings. 11th IEEE International Conference on*, pages 120–131. IEEE, 2003.

[159] V. Srinivasan and G. Varghese. Fast IP lookups using controlled prefix expansion. *ACM SIGMATICS*, 26(1):1–10, June 1998.

[160] Venkatachary Srinivasan, Subhash Suri, and George Varghese. Packet classification using tuple space search. In *ACM SIGCOMM Computer Communication Review*, volume 29, pages 135–146. ACM, 1999.

[161] Venkatachary Srinivasan and George Varghese. Fast address lookups using controlled prefix expansion. *ACM Transactions on Computer Systems (TOCS)*, 17(1):1–40, 1999.

[162] Venkatachary Srinivasan and George Varghese. A survey of recent ip lookup schemes. In *Protocols for High-Speed Networks VI*, pages 9–23. Springer, 2000.

[163] Venkatachary Srinivasan, George Varghese, Subhash Suri, and Marcel Waldvogel. *Fast and scalable layer four switching*, volume 28. ACM, 1998.

[164] Donpaul C Stephens and Hui Zhang. Implementing distributed packet fair queueing in a scalable switch architecture. In *INFOCOM'98. Seventeenth Annual Joint Conference of the IEEE Computer and Communications Societies. Proceedings. IEEE*, volume 1, pages 282–290. IEEE, 1998.

[165] Ivan Stojmenovic. Honeycomb networks: Topological properties and communication algorithms. *Parallel and Distributed Systems, IEEE Transactions on*, 8(10):1036–1042, 1997.

[166] Harold S Stone. Parallel processing with the perfect shuffle. *IEEE Transactions on Computers*, (2):153–161, 1971.

[167] Herbert Sullivan and Theodore R Bashkow. A large scale, homogeneous, fully distributed parallel machine, i. In *ACM SIGARCH Computer Architecture News*, volume 5, pages 105–117. ACM, 1977.

[168] L. Tassiulas. Linear complexity algorithms for maximum throughput in radio networks and input queued switches. In *INFOCOM '98. Seventeenth Annual Joint Conference of the IEEE Computer and Communications Societies. Proceedings. IEEE*, volume 2, pages 533–539 vol.2, Mar 1998.

[169] A. Thareja and A.K. Agrawala. On the design of optimal policy for sharing finite buffers. *Communications, IEEE Transactions on*, 32(6):737–740, Jun 1984.

[170] Ashok K Thareja and Ashok K Agrawala. Impact of buffer allocation policies on delays in message switching networks. In *INFOCOM*, pages 436–442, 1983.

[171] D. Tipper and M.K. Sundareshan. Adaptive policies for optimal buffer management in dynamic load environments. In *INFOCOM '88. Networks: Evolution or Revolution, Proceedings. Seventh Annual Joint Conference of the IEEE Computer and Communcations Societies, IEEE*, pages 535–544, March 1988.

[172] Brian Towles and William J Dally. Worst-case traffic for oblivious routing functions. In *Proceedings of the fourteenth annual ACM symposium on Parallel algorithms and architectures*, pages 1–8. ACM, 2002.

[173] Brian Towles and William J Dally. Guaranteed scheduling for switches with configuration overhead. *Networking, IEEE/ACM Transactions on*, 11(5):835–847, 2003.

[174] Leslie G Valiant and Gordon J Brebner. Universal schemes for parallel communication. In *Proceedings of the thirteenth annual ACM symposium on Theory of computing*, pages 263–277. ACM, 1981.

[175] George Varghese. *Network algorithmics*. Chapman & Hall/CRC, 2010.

[176] John Von Neumann. A certain zero-sum two-person game equivalent to the optimal assignment problem. *Contributions to the Theory of Games*, 2:5–12, 1953.

[177] Sigurd Waaben and P Carmody. High speed, high-current word-matrix using charge-storage diodes for rail selection. In *Proceedings of the December 9-11, 1968, fall joint computer conference, part II*, pages 981–986. ACM, 1968.

[178] Benjamin Wah. Wiley encyclopedia of computer science and engineering, 2008.

[179] Marcel Waldvogel, George Varghese, Jon Turner, and Bernhard Plattner. *Scalable high speed IP routing lookups*, volume 27. ACM, 1997.

[180] Guohui Wang, David G Andersen, Michael Kaminsky, Konstantina Papagiannaki, TS Ng, Michael Kozuch, and Michael Ryan. C-Through: Part-time optics in data centers. *ACM SIGCOMM Computer Communication Review*, 41(4):327–338, 2011.

[181] TING Wang, ZHIYANG Su, Y Xia, and MOUNIR Hamdi. Rethinking the data center networking: Architecture, network protocols, and resource sharing. 2014.

[182] S.X. Wei, E.J. Coyle, and M.-T.T. Hsiao. An optimal buffer management policy for high-performance packet switching. In *Global Telecommunications Conference, 1991. GLOBECOM '91. 'Countdown to the New Millennium. Featuring a Mini-Theme on: Personal Communications Services*, pages 924–928 vol.2, Dec 1991.

[183] Stewart Weiss. Parallel computing.

[184] Guo-Liang Wu and Jon W. Mark. A buffer allocation scheme for ATM networks: complete sharing based on virtual partition. *Networking, IEEE/ACM Transactions on*, 3(6):660–670, Dec 1995.

[185] Haitao Wu, Guohan Lu, Dan Li, Chuanxiong Guo, and Yongguang Zhang. MDCube: a high performance network structure for modular data center interconnection. In *Proceedings of the 5th international conference on Emerging networking experiments and technologies*, pages 25–36. ACM, 2009.

[186] Naoaki Yamanaka and Ken-ichi SATO. Performance limitation of leaky bucket algorithm for usage parameter control and bandwidth allocation methods. *IEICE Transactions on Communications*, 75(2):82–86, 1992.

[187] Kenji Yoshigoe and Kenneth J Christensen. A parallel-polled virtual output queued switch with a buffered crossbar. In *High Performance Switching and Routing, 2001 IEEE Workshop on*, pages 271–275. IEEE, 2001.

[188] DOI Yukihiro and Naoaki Yamanaka. A high-speed ATM switch with input and cross-point buffers. *IEICE Transactions on Communications*, 76(3):310–314, 1993.

[189] Francis Zane, Girija Narlikar, and Anindya Basu. Coolcams: Power-efficient TCAMs for forwarding engines. In *INFOCOM 2003. Twenty-Second Annual Joint Conference of the IEEE Computer and Communications. IEEE Societies*, volume 1, pages 42–52. IEEE, 2003.

[190] Yan Zhang and Nirwan Ansari. On architecture design, congestion notification, TCP incast and power consumption in data centers. *Communications Surveys & Tutorials, IEEE*, 15(1):39–64, 2013.

[191] Jin Zhao, Xinya Zhang, Xin Wang, Yangdong Deng, and Xiaoming Fu. Exploiting graphics processors for high-performance ip lookup in software routers. In *INFOCOM, 2011 Proceedings IEEE*, pages 301–305, April 2011.

Index